Astronomers' Universe

Series Editor

Martin Beech, Campion College, The University of Regina, Regina, Canada

The Astronomers' Universe series attracts scientifically curious readers with a passion for astronomy and its related fields. In this series, you will venture beyond the basics to gain a deeper understanding of the cosmos—all from the comfort of your chair.

Our books cover any and all topics related to the scientific study of the Universe and our place in it, exploring discoveries and theories in areas ranging from cosmology and astrophysics to planetary science and astrobiology.

This series bridges the gap between very basic popular science books and higher-level textbooks, providing rigorous, yet digestible forays for the intrepid lay reader. It goes beyond a beginner's level, introducing you to more complex concepts that will expand your knowledge of the cosmos. The books are written in a didactic and descriptive style, including basic mathematics where necessary.

William H. Waller

Crucibles of Creation

Exploring the Origins of Stars, Planets, and Life Within Galactic Ecosystems

Springer

William H. Waller
The Galactic Inquirer
Rockport, MA, USA

ISSN 1614-659X ISSN 2197-6651 (electronic)
Astronomers' Universe
ISBN 978-3-032-17257-0 ISBN 978-3-032-17258-7 (eBook)
https://doi.org/10.1007/978-3-032-17258-7

This Springer imprint is published by the registered company Springer Nature Switzerland AG
The registered company address is: Gewerbestrasse 11, 6330 Cham, Switzerland

Regions of lucid matter taking forms,
Brushes of fire, hazy gleams,
Clusters and beds of worlds, and bee-like swarms
Of suns, and starry streams.

—Alfred Tennyson in "The Palace of Art" (1833 CE)

To Tatiana Aleksandrovana Lozinskaya
(Sternberg Astronomical Institute, Moscow)
Pioneering researcher of massive stars and their environmental impacts

And to my wife Sandra Ellen Paille
Caring companion, confidante, and compass—in loving gratitude

Foreword

In the 13.8-billion-year chronology of the Universe following the Big Bang, the first stars and galaxies likely formed very early on—within roughly the first billion years. Stars continue to form today within gas-rich galaxies such as the Milky Way, so reading up on how stars and their planets form and develop over time can help you cover a lot of ground. In this book, author Bill Waller offers an environmental approach to getting current on the cosmic topic, one that recognizes interactions within and among the assemblages of stars, gas, dust, and other matter that are involved in the business of fostering new stars, planets, and the chemistry of life. Bill dubs these complexes *galactic ecosystems* to highlight their connectivity across a vast range of scales.

In the early 20th century, galaxies were first demonstrated to be celestial objects beyond our Milky Way. Since then, astronomy has advanced in leaps and bounds. The enterprise is now enabled by large observatories both on Earth and in space with capabilities that extend across the electromagnetic spectrum. These astrophysical observatories offer explorations of the universe across space and time with views that can be profound. Their pictures explore structures and components of both nearby star-forming nebulae and distant starbursting galaxies that are literally beyond anything we encounter in daily life. The images are inspirational, while understanding them can be anything but intuitive for us denizens of planet Earth.

Bill has experience in interpreting these views gained in the course of working as an astrophysicist at NASA's Goddard Space Flight Center, an astronomy professor at Tufts University, and a high school science teacher.

Online he is the co-founder and co-editor of *The Galactic Inquirer* and producer of the community-oriented *Doc Waller's Earth and Space Report* video series.

He is the co-author, with Paul Hodge, of *Galaxies and the Cosmic Frontier*, the author of *The Milky Way—An Insider's Guide*, *Astronomy—A Beginner's Guide*, and *An Earthling's Guide to the Solar System. Crucibles of Creation* is his latest insightful guide to astrophysical environments that can be astonishingly beautiful, stunningly violent, and cosmically nurturing.

Jerry Bonnell
Astronomy Picture of the Day
University of Maryland
College Park, USA

Preface

It is not hyperbole to state that galactic ecosystems have spawned pretty much everything that we hold dear—including ourselves. These nebular nurseries have incubated all the stars, planetary systems, and many of the complex molecules known to humanity. From there, the myriad of planets in orbit around their hosting stars have further refined the organic matter into even more complex biochemicals. On the surface of one particularly moist planet, life took hold, and evolutionary processes led to you and me collectively pondering the wonders of galactic ecosystems via this book.

The cosmic sweep involved in comprehending this tale is inevitably vast. On the microscale, we will encounter atomic nuclei and their alchemical transmutations within stars, the assembly of atoms into molecules within cryogenic clouds, and prospects for building up the macromolecules necessary for life as we know it. On much larger scales, we will explore protostars and protoplanetary disks inside molecular cloud cores a few light-years in size, their hosting molecular clouds tens of light-years across, giant molecular clouds and ionized nebulae hundreds of light-years in extent, and finally entire starburst galaxies spanning thousands of light-years. Along the way, we will gain an appreciation for the most massive, powerful, and short-lived stars as they transform their natal environs via their intense radiation, winds, and explosive deaths. Without the elemental beneficence of these stars, rocky planets, water, and life itself would be impossible.

I must admit to having nurtured over the decades a strong affinity for the galactic ecosystems that host hot massive stars, as their powerful ultraviolet radiation produces the stunning displays of ionized matter that grace many of the astronomy magazines and books that I grew up reading. Like a moth to a flame, I continue to be enthralled by these fluorescing realms and so have included a representative sampling for your own delectation.

Of course, these colorful renderings convey far more than their aesthetic appeal. By imaging galactic ecosystems at selected wavelengths across the electromagnetic spectrum, astronomers can diagnose their multi-phase content, structure, and powering. Many of these multi-wavelength images come from the most sensitive telescopes yet to be created by humans. These include at radio wavelengths—the *Very Large Array (VLA), Atacama Large Millimeter/submillimeter Array (ALMA), Submillimeter Array (SMA),* and *Planck Space Telescope*; at infrared wavelengths—the *Infrared Astronomy Satellite (IRAS), Infrared Space Observatory (ISO), Widefield Infrared Survey Explorer (WISE), Stratospheric Observatory for Infrared Astronomy (SOFIA), Spitzer Space Telescope, Herschel Space Observatory,* and *James Webb Space Telescope (JWST)*; at visible wavelengths—the *Gaia Astrometric Observatory, Hubble Space Telescope (HST), JWST* (at red wavelengths), and numerous ground-based optical observatories (including those operated by skilled amateur astronomers); at ultraviolet wavelengths—the *HST, Ultraviolet Imaging Telescope (UIT),* and *Galaxy Evolution Explorer (GALEX)*; and at X-rays—the *Chandra X-ray Observatory* and *XMM-Newton Observatory*. Considerable space in this book will be dedicated to the scientific "craft" of physically interpreting these multi-wavelength images.

Given the many celestial images that grace the pages of this book, I was delighted when Jerry Bonnell (University of Maryland) agreed to write the book's Foreword. Jerry is the co-founder of NASA's *Astronomy Picture of the Day (APOD)*—the go-to website for stunning visuals of the cosmos accompanied by well-informed descriptions (see https://apod.nasa.gov/apod/astropix.html). Together with Rob Nemiroff (Michigan Tech. U.), Jerry has hosted this vital resource for more than 30 years. On a more personal level, I have fond memories of my young family socializing with Jerry's family in the 1990s. These convivial gatherings occurred mostly at Jerry and Letty's welcoming home in Old Greenbelt, MD, just a short walk through the woods to NASA's Goddard Space Flight Center where we both worked. Those were precious times indeed!

I am grateful to the many other astronomers (present and past) whose insightful research and communications helped me in my own research and writing efforts. These include Lori Allen, John Bally, Francois Boulanger, Adeline Caulet, Mike Deneen, Mike Fanelli, Paul Goldsmith, Mark Hemeon-Heyer, Paul Hodge, Sue Kleinmann, Steve Kolaczkowski, Charlie

Lada, Charles Law, Eliot Malamuth, Pam Marcum, Brett McGuire, Eric Murphy, Norbert Shulz, Theo O'Neill, Phil Orbanes, Joel Parker, Travis Rector, Nick Scoville, Brooke Skelton, Steve Strom, David Tower, Christy Tremonti, Frank Varosi, Bill Wall, Tony Weinbeck, and Judy Young. These folks are in no way responsible for any errors that remain within the pages of this book. Those are mine alone to acknowledge and make amends.

My efforts to effectively convey the wonders of galactic ecosystems were greatly augmented by the graphic artwork of Leigh Slingluff. This is the fourth book of mine to benefit from her artistic expertise. Leigh's own artwork can be viewed at https://www.leighslingluff.com/.

The good folks at Springer Nature encouraged my initial conceptualizing of this book and gently guided the subsequent work through to the finished narrative that you see today. Thank you!

I hope that through reading this book, you will obtain a better understanding of your birthright as a "child of the cosmos." You share this birthright with all lifeforms on Earth and likely diverse other lifeforms on worlds yet to be explored. This sense of communion pervades the cosmos, made visceral through the awesome creativity of galactic ecosystems. Come join the journey!

Rockport, MA, USA William H. Waller

Competing Interests The author has no competing interests to declare that are relevant to the content of this manuscript.

Contents

1

Introduction

The Universe as we know it today is dominated by matter that is for the most part completely invisible. This so-called *dark matter* is thought to be structured into a vast web of intersecting filaments, whose nodes host the galaxy clusters and superclusters that we can observe with our telescopes. Evidence for the ponderous dark matter comes from its gravitational effects—from binding the motions of stars and gas clouds within individual galaxies to bending and distorting the light from more distant galaxies (Fig. 1.1).

The space that enfolds the dark web has been found to be expanding ever faster, taking its galaxian denizens on a wild ride with no end in sight. Cosmologists imagine this accelerated expansion being driven by some ubiquitous form of energy that remains utterly mysterious to the best of minds—what has been dubbed *dark energy*. Further support for overwhelming amounts of dark energy comes from details in the *Cosmic Microwave Background (CMB)* which require the Universe to have zero topological curvature and so be dependent on some sort of energy that can do the flattening. Together, the dark matter and dark energy have been estimated to comprise more than 95% of the total matter-energy in the current-epoch Universe (see Fig. 1.2).

The present book concerns itself with the remnant 5% of the cosmos that interacts with electromagnetic radiation and so can be observed. Most of this luminous material can be found in galaxies and their immediate surroundings. Here, myriad stars co-mingle with clouds of gas and dust in proportions depending on the evolutionary state of the galaxy. In most elliptical galaxies, the transformation of gas into stars is essentially complete, with very little gas remaining to form new stars. In many spiral and irregular galaxies, however,

W. H. Waller, *Crucibles of Creation*, Astronomers' Universe,
https://doi.org/10.1007/978-3-032-17258-7_1

Fig. 1.1 The galaxy cluster **Abell 370** as imaged by the Hubble Space Telescope reveals multiple distorted images of more remote galaxies, whose light has been gravitationally lensed by the foreground cluster's dark matter (*credits* NASA, ESA, and J. Lotz and the HFF Team [STScI])

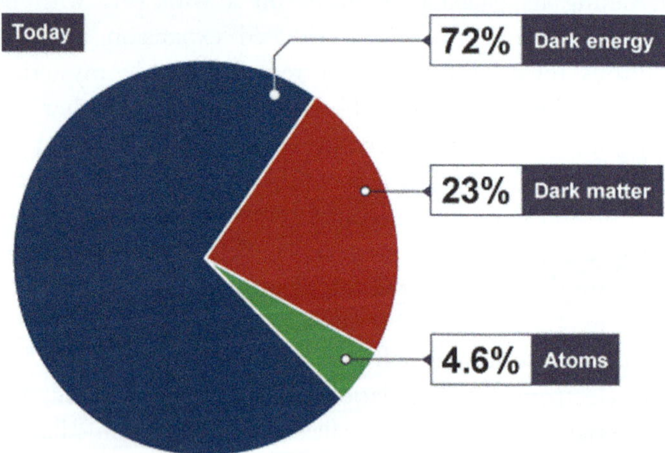

Fig. 1.2 The composition of matter-energy in the current-epoch Universe (*credit* BBC Education and NASA)

the transformative process continues apace, as gas clouds continue to gravitationally congeal and so spawn new generations of stars. This foment of celestial being and becoming forms the primary subject of the present book.

We will begin with our own Milky Way galaxy and what we can find here. Then, we will consider the nearest cold molecular clouds in terms of their structure and contents. The star- and planet-forming process is first introduced here. Already, we will be witness to fascinating tumult in the parental clouds that has been driven by jet-like outflows from the newborn stars. We will then widen our scope to include giant molecular clouds that have incubated clusters of stars big enough to include a few massive stars among the less massive hoard. And what a difference a massive star makes, as its intense ultraviolet radiation, powerful winds, and explosive deaths impact the natal cloud! Given enough massive stars in the cluster, the collective consequences can lead to rampant *starburst activity* that completely transforms the nebular nursery. We will examine a few good examples of the starbursting genre within the Milky Way and then move on to other powerful starbursts currently churning within nearby galaxies. Along the way, we will explore the biochemical character of these various *galactic ecosystems.*

For the most part, our story of nebular, stellar, and chemical transformations within galactic ecosystems will be told in the language of light. We will fully employ the dual interpretations of light as both particles and waves, being that both models are essential to explaining all that light does. The particle picture is most useful in understanding the creation and absorption of light. Excited atoms and molecules emit their excess energy in the form of photons. These particles of light (sometimes referred to as "wave packets") have distinct energies that relate directly to the specific frequencies of light that are emitted. The quantitative formula for this relation is

$$E = hf,$$

where E is the photon's energy, h is Planck's constant of quantum energetics, and f is the corresponding frequency. The wavelength, in turn, is related inversely to the frequency according to the "wave equation"

$$\lambda = c/f,$$

where λ is the wavelength and c is the speed of the light wave.

Given these fundamental relations, the emission lines of differing frequency (or wavelength) that are evident in spectra of molecular, atomic, and ionized nebulae can be diagnosed in terms of the gases emitting these photons and their corresponding states of excitation. The photon picture of light also is most helpful when understanding how light interacts with other atoms and molecules. Here, the photons are absorbed preferentially by matter that is primed to accept the corresponding energies. The absorption lines found in spectra of stars provides a key example of particular gases in the stellar atmospheres having absorbed photons at specific frequencies (see Appendix F). Another important function of light as a particle is when it is detected—by our retinas or by electronic sensors. Here, the process is best understood in terms of photons interacting with the matter in the detector on the quantum level.

The wave picture of light provides the best way to comprehend the propagation of light through space—and through our various telescopes and spectroscopes—on its way to being detected. The processes of light reflection, refraction, diffraction, interference, and polarization all can be quantitatively modeled by invoking electromagnetic waves of known frequencies, wavelengths, and amplitudes.

You might have noticed that both photons and waves involve wave action in some way. Photons can be conceptualized as wave packets, while electromagnetic waves can be understood as consisting of co-varying electric and magnetic fields that oscillate transverse to the overall direction of wave propagation. In this book, multiple wavelengths of light will be invoked, as different wavelength domains often indicate different physical processes operating within the galactic ecosystems (see Fig. 1.3). We currently live in a golden age of multi-wavelength astronomy, where all wavelengths can be sampled and analyzed—from nanometer-scale X-ray waves, to ever-longer ultraviolet, visible, infrared, and radio waves—the latter having wavelengths of centimeters to kilometers.

Equipped with our full quiver of light sensibilities, we are almost ready to forge ahead with exploring galactic ecosystems. But first we need to better understand what is meant by such an evocative term, the subject of the next chapter.

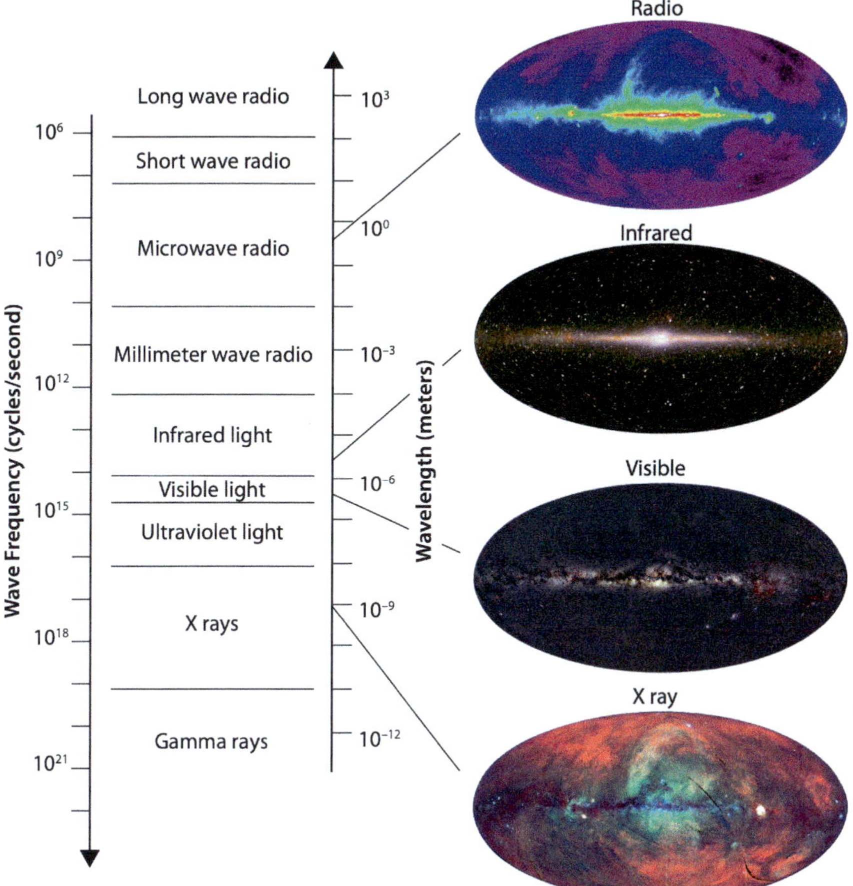

Fig. 1.3 The electromagnetic spectrum with all-sky views at representative wavelengths and corresponding frequencies (*credits* L. H. Slingluff, with the all-sky images from the Multiwavelength Milky Way website hosted by NASA's Goddard Space Flight Center at asd.gsfc.nasa.gov/archive/mwmw/mmw_allsky.html)

Part I

Galactic Ecosystems in Context

Within the clouds of dust and light,
Swirling veils obscure the sight,
Yet stars are forged from endless night.

—Michael Patrick Deneen and ChatGPT 3.5 (2025 CE)

2

Defining Galactic Ecosystems

The term *galactic ecosystem* figures large in this book, yet it remains mostly unused in the scientific and lay literature. Exceptions include a few papers and conference proceedings, several articles by radio astronomer Michael Burton promoting the term, and a chapter in the technical book by astrophysicist Xander Thielens on *The Physics and Chemistry of the Interstellar Medium* (Cambridge University Press 2005). I have chosen to advance the salience of *galactic ecosystems*, as I believe such entities best encapsulate the cosmic dramas that unfold within their domains. Other terms related to galactic ecosystems include the interstellar medium (ISM), dark clouds, molecular clouds (MCs), star-forming regions, stellar nurseries, emission nebulae, ionized nebulae—aka HII regions, giant HII regions (GHRs), starbursts (SBs), etc. But none of them tell enough of what's going on. They especially don't capture the consequential interactions among the various stellar and nebular components. To further understand galactic ecosystems in their own right, let's parse the term word by word. A typical dictionary (such as *Miriam Webster's New Collegiate Dictionary* used here) would give something like the following definitions …

Galactic: Of or relating to a galaxy, esp. the Milky Way.

Ecosystem: The complex of a community and its environment functioning as an ecological unit in nature.

Ecology: The totality or pattern of relations between organisms and their environment.

W. H. Waller, *Crucibles of Creation*, Astronomers' Universe,
https://doi.org/10.1007/978-3-032-17258-7_2

Fig. 2.1 *Left* The Atacama Pathfinder Experiment (*APEX*) which operates at millimeter and submillimeter wavelengths in the high Chilean desert (*credits* European Southern Observatory, Babak Tafreshi). *Right* A portion of the nearby **Taurus Molecular Cloud** as imaged by *APEX* at a wavelength of 870 μm (microns). The image spans 7 light-years. It shows glowing dust entrained within a filamentary cloud along with several dense star-forming cores of dust and molecular gas within the filament (*credits* SO/APEX [MPIfR/ESO/OSO]/A. Hacar et al./Digitized Sky Survey 2. Acknowledgment: Davide De Martin)

This latter definition refers specifically to living organisms, but contains important verbiage on relations that can be expanded to the following working definition …

Galactic Ecosystem: A complex of inter-relating stellar, nebular, and other matter that is actively forming new stars. Together, the components comprise a functioning environmental unit while responding to even larger environmental influences (see Fig. 2.1). Over its lifetime, such a complex is critically vulnerable to any newborn massive stars that it incubates.

In keeping with all the other related terms and their abbreviations, I suppose it is appropriate to give galactic ecosystems their own abbreviation, namely GEs. Whether this acronym will gain significant favor within the astronomical literature remains to be seen.

As theaters of cosmic transformation, galactic ecosystems encompass a great range of spatial scales … from molecular cloud cores (0.1–1.0 light-years) to nebular pillars (1.0–10.0 light-years), HII regions (10.0–1000 light-years), giant molecular clouds (~ 100 light-years), spiral arms (~ 1000 light-years), galactic winds (~ 10,000 light-years), and even interacting galaxies (~ 100,000 light-years). This impressive range provides an essential link between the much smaller realm of individual stars and their planetary systems (light-seconds to light-hours) versus the far more cosmic realm that ranges from small galaxy groups to large-scale superclusters and filaments spanning millions to billions of light-years.

Such fecund regions also play host to energetic *stellar feedback* which can lead to major transformations of the nebular structures along with new forms of nebular chemistries—up to and including ***biochemicals***. For the sake of clarity, definitions of these respective terms follow.

Stellar Feedback: Ultraviolet radiation, powerful winds, and supernova blasts from newborn massive stars can transform the architecture of galactic ecosystems while transmuting their chemical makeup.

Biochemicals: The molecules that play key roles in sustaining life on Earth and perhaps elsewhere—e.g. H_2O, H_2CO, O_2, CH_4, HCN, NH_3, PO_4, amino acids, nucleotides, etc.

After attending a Zoom presentation that I gave on galactic ecosystems, my polymath cousin, David Kurt, wondered whether they could be likened to the microbial ecosystems that inhabit our guts. There, all sorts of interactions occur that have important consequences for our overall health. After thinking this over offline, I would have to agree that both systems involve important molecular interactions. However, I don't think that galactic ecosystems share the same stability, or ***homeostasis***, that characterizes healthy guts. Instead, galactic ecosystems are rapid evolvers, drastically transmogrifying in the presence of the stars that they have spawned. Timescales for maintaining their physical integrity can be as short as a few tens of millions of years. Compared to the age of Earth and the rest of our Solar System, such timescales are the equivalent of one day in a year. Indeed, we don't know which galactic ecosystem produced the likes of us, as our birth cloud has long since dissipated into the larger galactic disk.

Given this level of impermanence, perhaps a better analogy would be the transient thermal pools at Yellowstone National Park which are renowned for fostering diverse and evolving microbial communities during their brief lifetimes. Another, less edifying analogy, would be the natural environments that get disrupted and often destroyed in the presence of the apex species that they have nurtured. We humans are currently prime examples of this home-wrecking analogy.

In 2023, I asked my friend Michael Deneen, a professor of English Literature, if he could work with a generative artificial intelligence program to provide a one-paragraph description of galactic ecosystems. Here is what Chat GPT-3.5 had to say, when Michael queried it …

Galactic ecosystems refer to the complex web of interstellar and planetary interactions within a galaxy, where celestial bodies, such as stars, planets, asteroids, and gas clouds, play interconnected roles in shaping the environment and conditions for life. These ecosystems encompass a wide range of phenomena, from the birth and death of stars, which release elements essential for life into space, to the gravitational interactions that influence the formation and stability of planetary systems. Within these systems, the emergence and evolution of life on suitable planets further contribute to the intricacies of galactic ecosystems, as life can impact its environment and, in some cases, even extend beyond its home planet to explore and adapt to the broader cosmic surroundings. Understanding the dynamics of galactic ecosystems is crucial not only for studying the origins and distribution of life in the universe but also for comprehending the long-term evolution and sustainability of cosmic environments.

If I were to grade this description, I would give it a B-. It contains several apt phrases, including "complex web of … interactions," "birth and death of stars," and "shaping … conditions for life." But then, it goes off the deep end in having life "impact its environment." Considering the range of spatial scales spanned by galactic ecosystems, I have difficulty imagining that life on any planet could have much of a physical effect over these vast distances. Perhaps some planetary microbes can join the overall interstellar mix over hundreds of millions of years in a process known as *panspermia*, but I wouldn't hold my breath. I can imagine, however, intelligent and technologically capable lifeforms communicating information across the expanse. Such a communication might include the haiku that Michael prompted Chat GPT-3.5 to generate regarding galactic ecosystems. Here's what he got …

Cosmic dance unfolds,
Stars and planets, life's embrace,
Galactic secrets.

This one is much harder to grade, but it seems to evoke an enfolding and mysterious agency to galactic ecosystems which I find rather endearing. What do you think?

In the next chapter, we will explore the spatial distribution of galactic ecosystems within the Milky Way. Here, multiple wavelengths are invoked to trace both the obscuring and emitting effects of the dust and gas that inhabit galactic ecosystems and hence betray their whereabouts.

3

Delineating Galactic Ecosystems

Now that we have some idea of galactic ecosystems as nebular complexes that host energetic and impactful star-forming activity, how do we go about finding them? Thanks to mapping surveys at multiple wavelengths, we now have reliable indications of their distribution throughout the Milky Way galaxy. Beyond our home galaxy, we can survey galactic ecosystems churning in many other star-forming galaxies. But let's get back to the Milky Way galaxy, where we can begin by comparing all-sky mosaics at visible and far-infrared wavelengths (see Fig. 3.1).

We can see that the visible-light mapping is sensitive to the presence of nearby clouds of obscuring dust that are seen in silhouette against the stellar background. This dust closely matches maps of nearby cold molecular clouds as traced by millimeter-wave emission from the carbon dioxide (CO) resident within these clouds. The clouds themselves are mostly made of molecular hydrogen (H_2). However, this diatomic gas is mostly invisible unless excited by strong shock waves or intense ultraviolet light. That is why astronomers have often utilized the obscuration evident at visible wavelengths to create maps of extinction and corresponding masses of dust and associated gas for the nearest clouds. Extinction in the Chameleon, Taurus and Perseus molecular clouds among others have been mapped this way (see Fig. 3.1a).

Interspersed amongst the starlight and obscuring clouds, reddish patches declare the presence of fluorescing hydrogen atoms that have been photoionized by the extreme ultraviolet emission coming from caches of newborn hot stars. The red glow arises during the recombination process—as recaptured electrons cascade from the 3rd to 2nd energy levels in the reconstituted atoms. These so-called HII regions trace the most recent star-forming activity

W. H. Waller, *Crucibles of Creation*, Astronomers' Universe,
https://doi.org/10.1007/978-3-032-17258-7_3

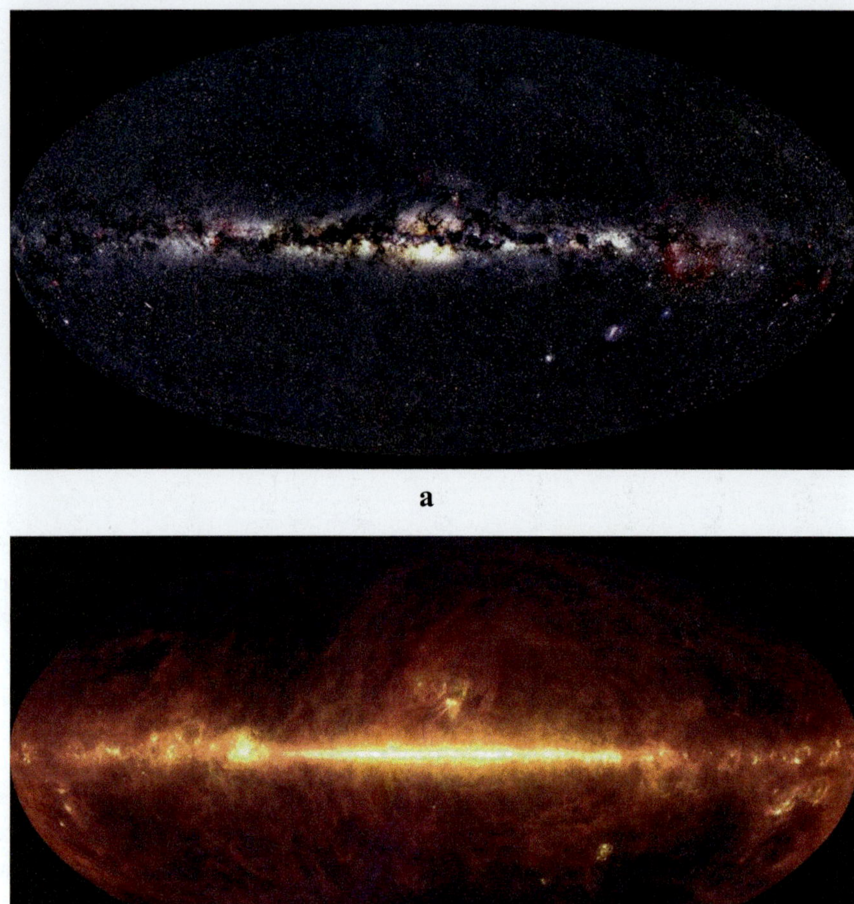

Fig. 3.1 **a** All-sky mosaic at visible wavelengths, featuring luminous stars, red fluorescing nebulae, and obscuring clouds of gas and dust along the plane of the Milky Way. The obscuring clouds represent relatively nearby galactic ecosystems seen in silhouette against the stellar background (*credit* Courtesy of Axel Mellinger, Central Michigan University). **b** All-sky mosaic at far-infrared wavelengths of 25, 60, and 100 μm, featuring glowing clouds of star-warmed dust along the plane of the Milky Way. The dust is responding to the energizing presence of newborn stars within the myriad galactic ecosystems that inhabit the galactic disk (*credits Infrared Astronomy Satellite [IRAS]*, IPAC, NASA)

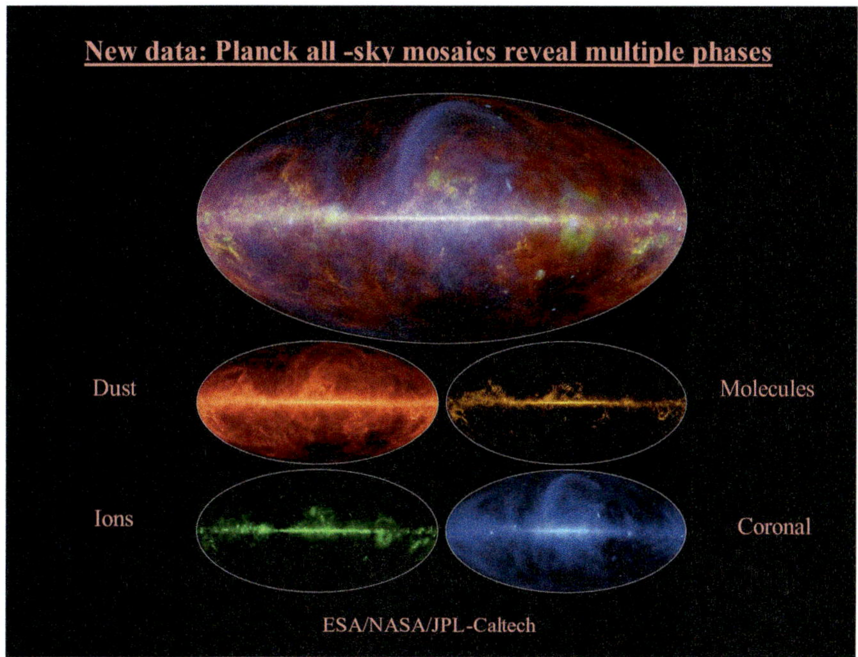

Fig. 3.2 Multi-wavelength all-sky mosaics from the *Planck* space observatory have been modeled according to the different emitting phases of the interstellar medium—including the star-warmed granular dust, cryogenic molecules at temperatures of a few Kelvins, "warm" ions at temperatures of several thousand Kelvins, and even hotter "coronal" plasma at temperatures of several million Kelvins (*credits Planck Space Telescope*, ESA/NASA/JPL-Caltech)

in those parts of the Milky Way galaxy that show minimal obscuration. By contrast, the far-infrared mapping traces star-warmed clouds of glowing dust distributed throughout the Milky Way (see Fig. 3.1b). The dust itself is thought to be very fine, akin to that of cigarette smoke. The far-IR mapping is most sensitive to dust grains with temperatures of several tens of degrees Kelvin.

More recent mappings by the *Planck* space observatory have yielded a cornucopia of images spanning a range of wavelengths from 300 μm in the far-infrared to 11,000 μm (11 mm) in the radio. Subsequent analysis of this multi-wavelength harvest has enabled scientists to pick apart the major emitting contributors (see Fig. 3.2). These include star-warmed dust grains, cryogenic molecules, warm ions, and hot (coronal) gas. These *Planck* decompositions are truly revelatory and bear much greater scrutiny.

Meanwhile, a lot can be accomplished by focusing on the visible and far-infrared mappings that have been available for many years. This is what my colleagues and I have done in the far-infrared by accessing the free space data utility *Skyview* which is curated at https://skyview.gsfc.nasa.gov/current/cgi/titlepage.pl. We used the *Skyview* Query Form to access data sets from the *Infrared Astronomical Satellite (IRAS)* and construct large mosaics of the 100 μm wavelength emission in galactic coordinates. The mosaics shown here each span 100° in galactic longitude and 60° in latitude with respective centers of 0°, 90°, 180°, and 270° of longitude—with 5° overlaps at each end. These 0.05° resolution mosaics were then spatially filtered to remove the strong brightening towards the galactic midplane. The result was a "froth" of far-infrared-emitting dust clouds along with a concentration of bright sources close to the plane of the Milky Way. Herein can be found rich menageries of galactic ecosystems (see Fig. 3.3).

For the far-infrared mosaics from the *IRAS* mission, the process of dividing the original mosaic by the median-smoothed mosaic leads to a mapping of relative enhancements and diminutions in the emission, with the strong gradient toward the galactic midplane effectively removed (see Fig. 3.3). Structuring in the residual emission is evident at all latitudes, the consequence of the structured emission scaling with the smooth emission that was divided out. This behavior is indicative of looking through a disk that contains both nearby and more distant sources of radiating dust. If the median-normalized structuring was only due to a nearby scrim of galactic "cirrus," then it would peter out close to the midplane. Some of the structuring has filamentary morphologies; other structuring is evocative of arcs and shell fragments. These latter features may denote nebular relics from prior star-birth and star-death activity. Then there are the discrete emitting regions which denote active galactic ecosystems.

We have analyzed the distribution of structuring in the median-normalized mosaics and find that they follow a power law of structuring with respect to spatial wavelength or frequency (see Fig. 3.4). The overall trend is for ever-decreasing power of the structuring at smaller spacings. There are kinks in the power-law slope at spacings of about 1° and 0.2°, indicative of changes in the structuring itself—from anecdotal morphologies to a turbulent froth, and finally a haze due to degrading resolution. The self-similar nature of these statistics can be interpreted in the parlance of fractal structuring. For example, looking towards the Galactic Center, the "turbulent" structuring has a power-law slope of beta = − 3.05 with a corresponding fractal dimension of D = 2.48. For reference, a fractal dimension of D = 2 would denote a smoothly varying or randomly varying surface, while a fractal dimension of D = 3 would indicate even more complex structuring than is apparent here.

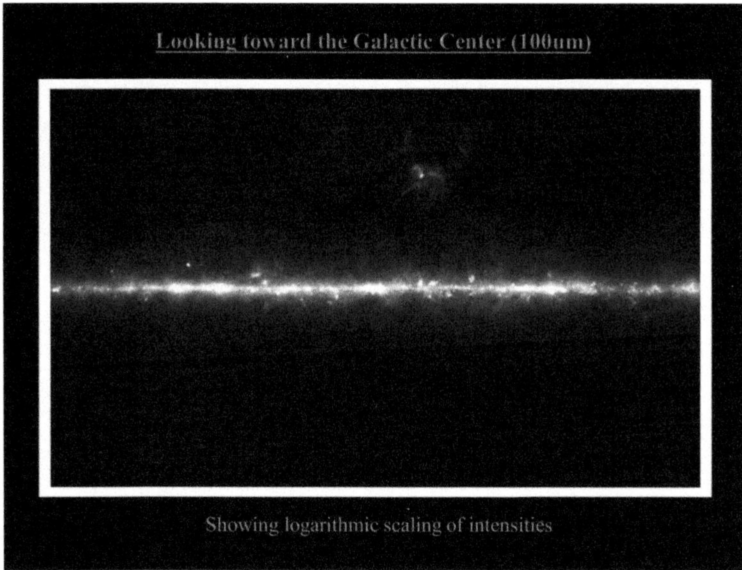

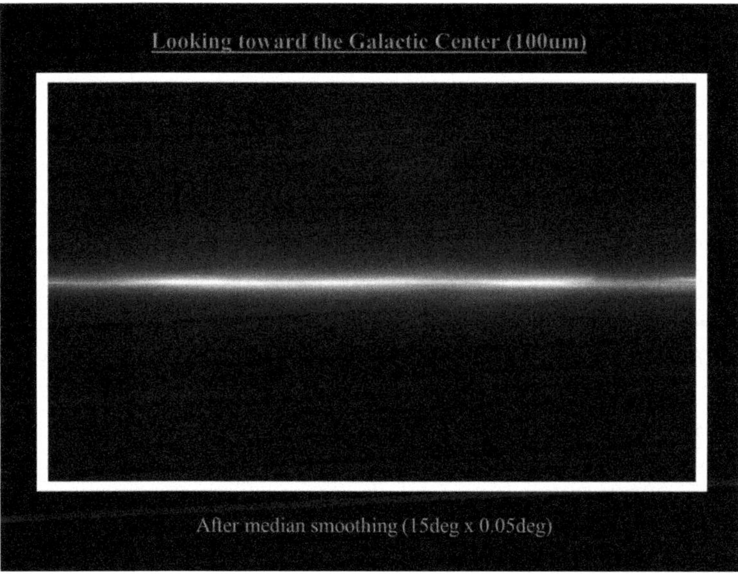

Fig. 3.3 100° by 60° mosaics of 100 μm emission from star-warmed dust in the direction of the Galactic Center. (Top) original mosaic with logarithmic scaling of intensities, (Middle) median smoothed mosaic of emission, (Bottom) spatially filtered mosaic, after dividing the original mosaic by the smoothed mosaic. Along the midplane, numerous bright galactic ecosystems are evident clear to the far-side of the Galaxy (*credits IRAS*/IPAC/JPL/NASA, Waller, W. H. et al. 1996)

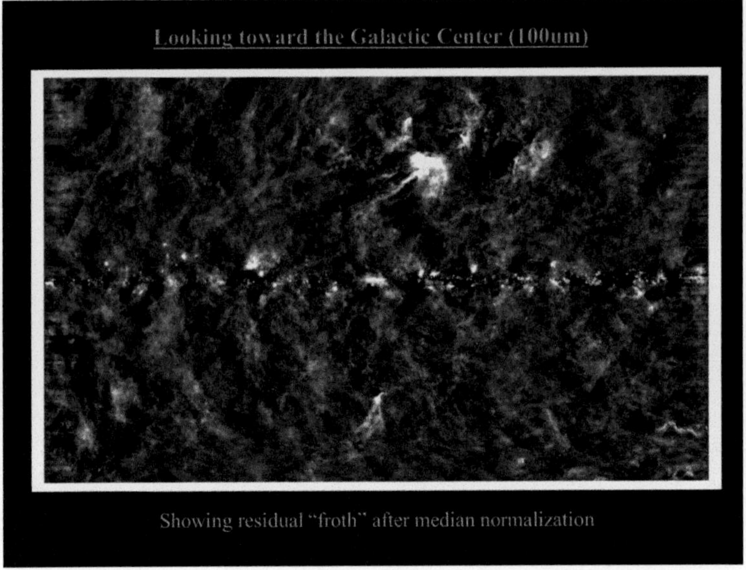

Fig. 3.3 (continued)

These statistics in the median-normalized structuring of far-infrared emission remain consistent all around the Milky Way. Even the outer Galaxy centered at a longitude of 180° shows similar spatial power law slopes and dimensions. That tells us the Galaxy is suffused with an interstellar medium of heterogeneous yet statistically uniform character.

These mosaics of continuum FIR emission have no kinematic information (unlike surveys of the HI and CO line emission) or other data that could provide 3-D renderings of the emitting matter. Nevertheless, even in projection, the statistics of structuring are consistent with a turbulent origin—as has been found in CO emission-line studies of the kinematics in nearby molecular clouds at high galactic latitude. Correlations with CO, HI, and radio-continuum surveys show that most of the "froth" can be related to clouds of atomic hydrogen (HI) emitting at 21 cm wavelength (49%), followed by radio-continuum emission from ionized gas (32%), and CO emission from molecular gas (19%). The CO correlations likely were underestimated due to the lower sensitivity of the CO surveys available at the time. The more discrete galactic ecosystems often show correlations with all the molecular, atomic, and ionized gas phases. To better ascertain the whereabouts of these galactic ecosystems, we will now examine both the visible-light and spatially-filtered FIR views of the Milky Way more closely (see Figs. 3.5, 3.6, 3.7 and 3.8).

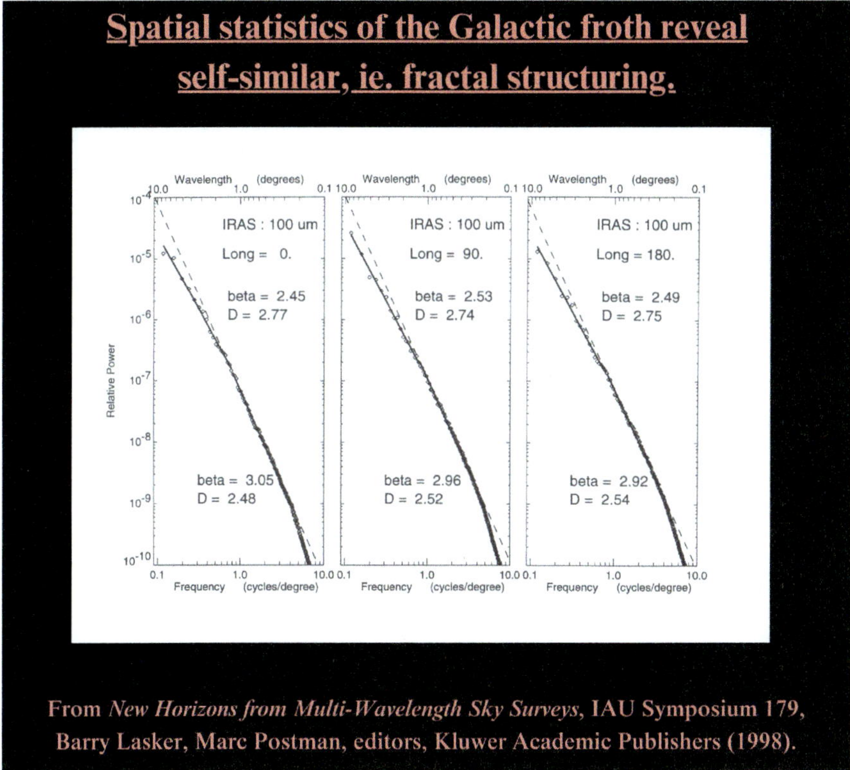

Fig. 3.4 Distribution of structural powering in median-normalized mosaics of the far-infrared emission in the Milky Way. The x-axis has spatial frequency increasing to the right and spacing ("wavelength") decreasing accordingly. The structural powering follows a power law for the most part—decreasing at higher spatial frequency and smaller spacing (*credits* W. H. Waller et al. 1998, IAU Symp. 179)

Looking Towards the Galactic Center

In the visible-light image of Fig. 3.5, dark clouds of gas and obscuring dust can be seen at nearly all longitudes. From right to left, these relatively nearby clouds of star-forming potential extend from the constellation of Centaurus through Norma, Lupus, Scorpius, Ophiuchus (including the Rho Ophiuchi dark cloud), Sagittarius (including the Pipe and Dark River Nebulae), Serpens, Scutum, and Aquila (including the Aquila Rift).

Among these dark clouds, several seemingly smaller regions of luminous gas are evident glowing in the red light of ionized hydrogen. Because we are looking towards the inner Galaxy here, we are seeing HII regions that typically belong to the Sagittarius spiral arm approximately 4500 light years away. Therefore, they have actual dimensions that are much larger than they appear

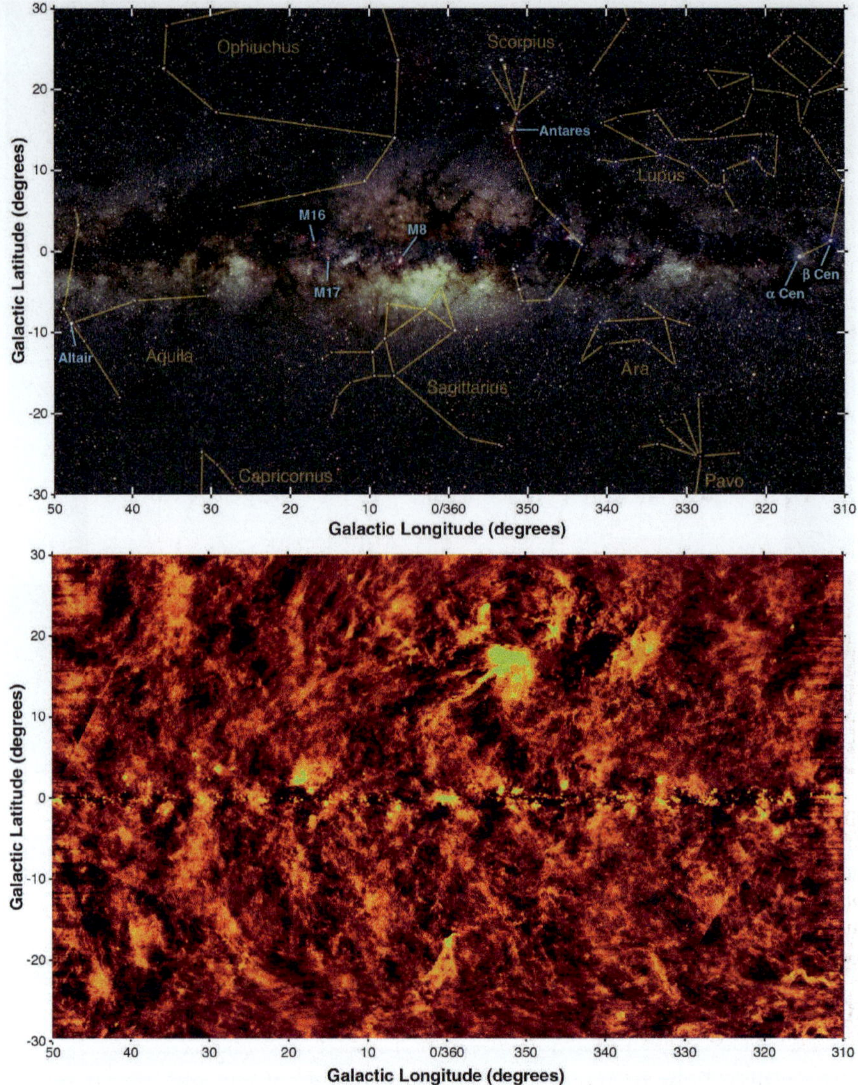

Fig. 3.5 (Top) Visible-light view of the Milky Way in the direction of the Galactic Center with constellations and prominent objects labeled (*credits* Axel Mellinger, Central Michigan University, W. H. Waller 2013). (Bottom) Median-normalized view of this region at 100 μmwavelength (*credits IRAS*/IPAC/JPL/NASA, W. H. Waller et al. 1996, L. H. Slingluff)

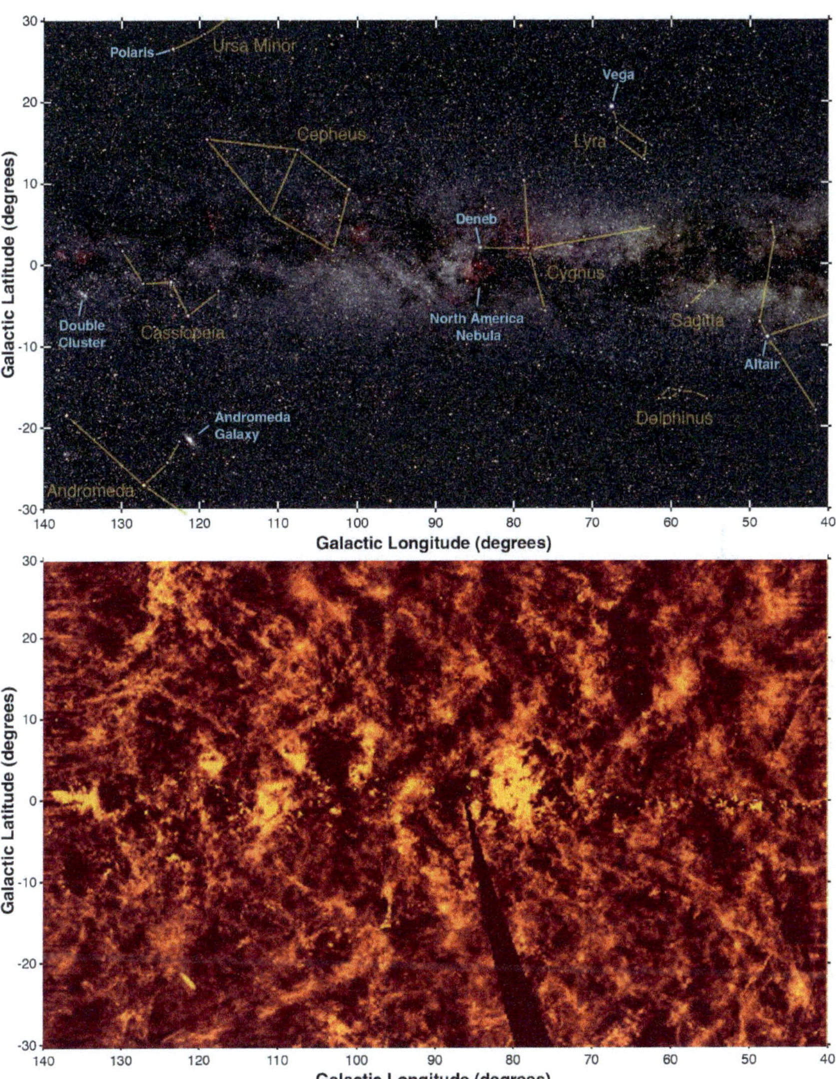

Fig. 3.6 (Top) Visible-light view of the Milky Way in the direction of the Sun's orbital motion, with constellations and prominent objects labeled (*credits* Axel Mellinger, Central Michigan University, W. H. Waller 2013). (Bottom) Median-normalized view of this region at 100 μmwavelength (*credits*: *IRAS*/IPAC/JPL/NASA, W. H. Waller et al. 1996, L. H. Slingluff)

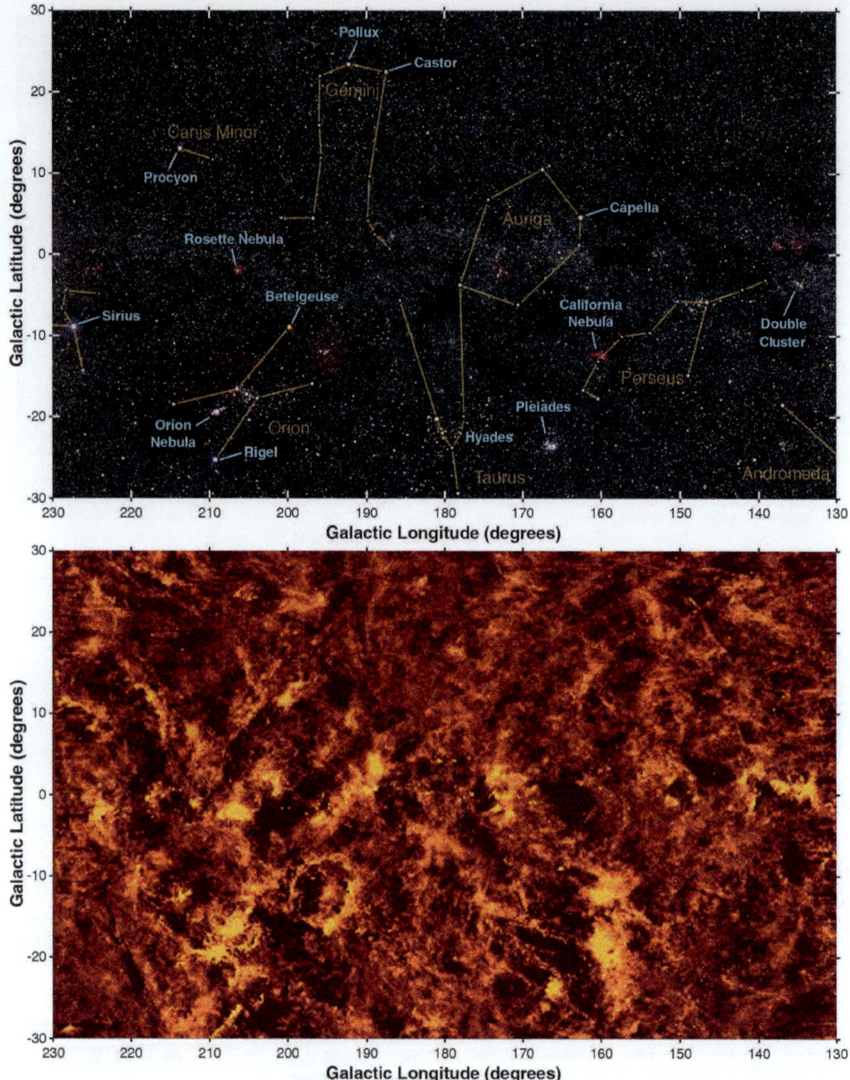

Fig. 3.7 (Top) Visible-light view of the Milky Way in the direction opposite the Galactic Center with constellations and prominent objects labeled (*credits* Axel Mellinger, Central Michigan University, W. H. Waller 2013). (Bottom) Median-normalized view of this region at 100 μmwavelength (*credits IRAS*/IPAC/JPL/NASA, W. H. Waller et al. 1996, L. H. Slingluff)

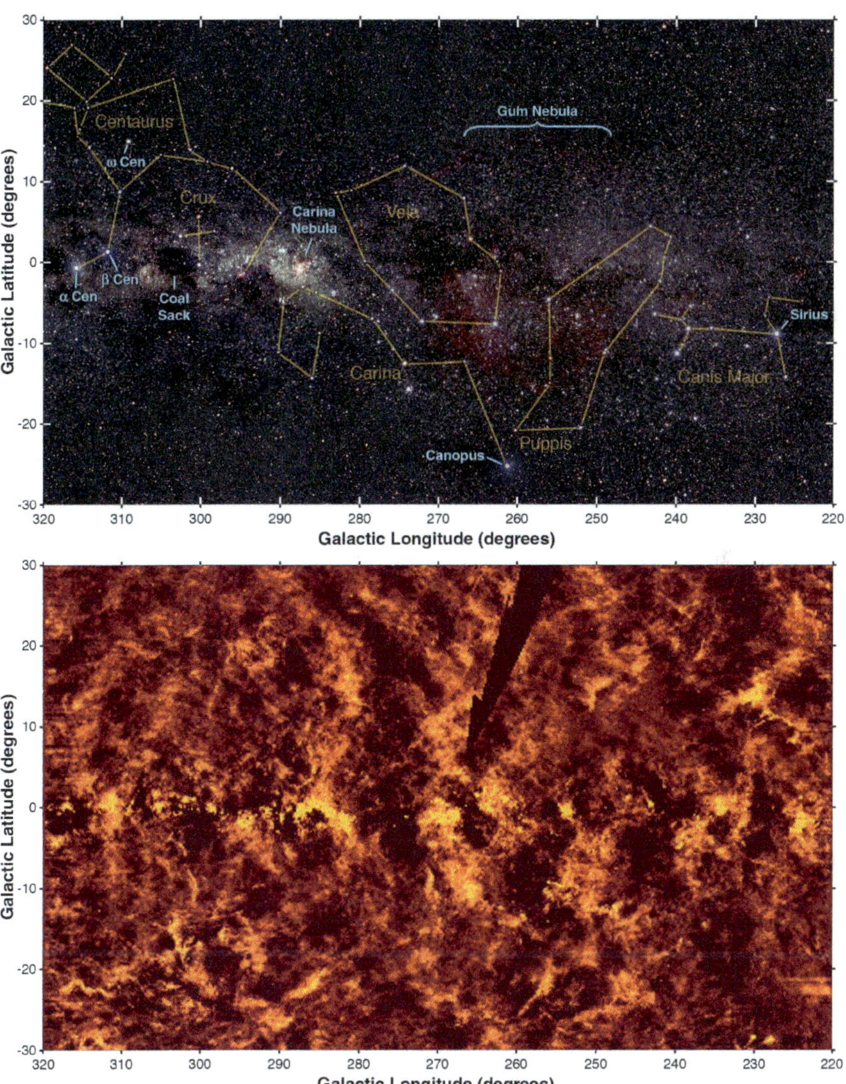

Fig. 3.8 (Top) Visible-light view of the Milky Way in the direction opposite the Sun's orbital motion with constellations and prominent objects labeled (*credits* Axel Mellinger, Central Michigan University, W. H. Waller 2013). (Bottom) Median-normalized view of this region at 100 μmwavelength (*credits IRAS*/IPAC/JPL/NASA, W. H. Waller et al. 1996, L. H. Slingluff)

to be—spanning tens to hundreds of light years. Each of these ionized regions is powered by the extreme ultraviolet light from collections of hot newborn stars. Notable HII regions in this region are listed in Table 3.1, where "NGC" refers to the *New General Catalogue of Nebulae and Clusters of Stars* begun by William and Caroline Herschel in the nineteenth century. "M" refers to the earlier catalog of nebulous looking objects compiled by Charles Messier in the eighteenth century. All data are from Richard Powell's online *Atlas of the Universe* (2006).

The median-normalized image at 100 μm wavelength reveals a far greater number of discrete emitting sources than are evident in the visible-light image. Indeed, in this swath of the Milky Way alone, more than 100 FIR-emitting sources have been cataloged (Waller et al. 1996). This disparity between the visible-light and FIR source counts results from the extensive population of obscuring dust clouds towards the inner Galaxy; the clouds' ubiquity reducing the number of visible HII regions that are evident from our perspective. The *Wide-field Infrared Survey Explorer (WISE)*, with its higher-resolution mapping capability along the Milky Way, has since increased the number of known sources in this region to more than 1000 (Anderson et al. 2014). However, the longer-wavelength *IRAS* mapping shown here can detect dust features at much cooler temperatures, thus revealing.

- a shell fragment at (G336-23) spanning 8°,
- the Lupus star-forming region at (G334+18),
- the Rho Ophiuchi star-forming region at (G353+17),
- another shell fragment at (G350+22) spanning 10° where second-generation star-forming activity is evident along its rim,

Table 3.1 HII regions in the direction of the Galactic Center (from Powell 2006)

Catalog name(s)	Common name	Longitude	Latitude	Distance/size (ly)
NGC 6188		337.2	− 1.1	4400/25
NGC 6334	(Cat's Paw Nebula)	351.1	+ 0.6	5500/60
NGC 6357		353.2	+ 1.0	3300/90
NGC 6514/M20	(Trifid Nebula)	7.0	− 0.3	4000/35
NGC 6523/M8	(Lagoon Nebula)	6.0	− 1.2	4500/100
NGC 6618/M17	(Omega/Swan Nebula)	15.1	− 0.8	4500/50
NGC 6611/M16	(Eagle Nebula)	17.0	+ 0.8	5700/60
Sharpless 2-54	(Serpent Nebula)	18.6	+ 1.9	6200/150

- the Corona Australis molecular cloud at (G360-18) whose FIR-emitting "tail" appears shaped by some external agent(s),
- an extended source associated with the Serpent Nebula at (G18+2),
- and hints of yet another shell fragment at (G32-18) that spans 14°.

Looking in the Direction of the Sun's Orbital Motion

Pointing toward the Sun's direction of orbital motion around the Galaxy, the visible-light mosaic continues the theme of dark dust clouds (or rifts) in silhouette against the starry background. These clouds are most pronounced in the directions of Aquila, Vulpecula, Cygnus, Cepheus, and Cassiopeia. Prominent HII regions trace the Orion spiral arm, of which we are a part. These are listed in Table 3.2.

The map of median-normalized FIR emission shows these sources plus a whole lot more. Beginning at a Galactic longitude of 40°, we can see a multitude of discrete sources close to the midplane. These likely represent galactic

Table 3.2 HII regions in the direction of the Sun's orbital motion (from Powell 2006)

Catalog name(s)	Common name	Longitude	Latitude	Distance/size (ly)
NGC 6820	(Pillars in Vulpecula)	59.4	− 0.2	6200/40
NGC 6888	(Crescent Nebula)	75.5	+ 2.4	4700/25
IC 1318	(Butterfly Nebula)	78.5	+ 1.7	4000/60
NGC 7000	(N. America Nebula)	85.5	− 1.0	1900/70
IC 1396	(Elephant's Trunk)	98.4	+ 4.6	2400/130
Sharpless 2-190	(Heart Nebula)	134.4	+ 0.8	6000/170
Sharpless 2-199	(Soul Nebula)	137.6	+ 1.2	6500/200

Notes "IC" refers to the *Index Catalogue of Nebulae and Clusters of Stars* published by John Dreyer in 1895 as a supplement to the NGC catalog. The Crescent Nebula (NGC 6888) is not actually a classic HII region consisting of natal gas that has been photoionized by newborn hot stars. Instead, it is an emission nebula whose gas was produced by powerful winds from a single hot Wolf-Rayet star. Also, the Heart and Soul nebulae reside considerably farther away than the others, being in a part of the more distant Perseus spiral arm ecosystems within the molecular ring at a galactocentric radius of 16,000 ly—60% of the distance from the Galactic Center to the Solar System. A bright duo of FIR spots shines at (G49.1-0.5) which could be associated with the HII region Sharpless 2-79. But the main attraction here is the Cygnus star-forming complex at (G78+2). The North America Nebula (NGC 7000) represents only the unobscured nearside of this prodigious galactic ecosystem. Here, star-warmed dust profusely radiates over a region spanning at least 10°. The estimated distance of the Cygnus complex is about 4600 ly which implies a physical extent of at least 800 ly. Sometimes dubbed Cygnus X due to the early detection of bright radio emission in this part of the sky, the Cygnus complex plays host to all sorts of intense star-forming activity—what could be justifiably called starburst activity. These hallmarks include millions of solar masses worth of molecular gas, rich associations of newborn hot O and B-type stars, hundreds of individual HII regions, and an X-ray emitting "superbubble" of hot gas that was likely blown by recent supernova activity. This one region takes the award as the most powerful galactic ecosystem within 7000 light-years of the Sun.

Beyond Cygnus, an impressive shell of FIR-emitting dust takes up much of the Cepheus constellation. This object, known as the Cepheus Bubble, is located at (G103+7) and spans about 11°, or 460 ly at an estimated distance of 2400 ly. It is associated with the Elephant's Trunk Nebula (IC 1396) which occupies its rightmost rim. Here, we have a compelling example of secondary star-forming activity that has followed a major blowout from a prior episode of massive star formation. Another shell-like morphology is evident at (G118+6), where the angular extent of 3° corresponds to a physical dimension of 140 ly at a distance of 2600 ly. Designated NGC 7822, the region encompasses the HII region Sharpless 2-171 which is being photoionized by the Cepheus OB4 cluster of young hot stars within. Like the Cepheus Bubble, NGC 7822 has an FIR morphology indicative of secondary star-forming activity on its rim.

Looking Opposite the Galactic Center

Looking in the direction opposite the Galactic Center, we can see that the starlight along the Galactic plane is greatly diminished—the consequence of viewing the sparsely populated "outback" of our home Galaxy. Even so, dark tendrils of obscuring dust can be identified coursing through the constellations of Perseus, Auriga, and Taurus. These dark clouds represent some of the nearest molecular clouds to us, the subject of the next chapter. Interspersed among the dark nebulosity, several roseate HII regions announce pockets of recent massive starbirth activity. These include the California Nebula in Perseus, the Rosette Nebula in Monoceros, and the Orion star-forming complex. These and other HII regions are listed in Table 3.3.

Many of these HII regions are situated 1300–1500 ly from us, in the main part of the Orion spiral arm. The notable exceptions in this listing are thought to be tracing parts of the more distant Perseus and Cygnus spiral arms.

Even though our views in this direction are not nearly as plagued by foreground clouds of obscuring dust, the far-infrared mapping shows that much remains hidden from visible reconnaissance. For example, the Perseus star-forming region comes alive with luminous activity, while the Orion complex positively flaunts its transformative dynamics. Highlights include

- In Perseus, IC 438 at (G160-18) shines as brightly in the far-infrared as NGC 1499 (the California Nebula), yet it contains no ionizing stars (see Appendix A.2).
- In Taurus, the Pleiades star cluster (M45) at (G167-25) shows prominently in the FIR, as the cluster is passing through an ambient cloud of gas and dust.

Table 3.3 HII Regions in the direction opposite the Galactic Center (from Powell 2006)

Catalog name(s)	Common name	Longitude	Latitude	Distance/size (ly)
NGC 1499	(California Nebula)	160.1	− 12.3	1500/70
IC 405	(Flaming Star Neb.)	172.1	− 2.3	1700/15
IC 410	(Tadpole Nebula)	173.6	− 1.8	11,000/60
NGC 1931	(Fly Nebula)	173.9	+ 0.3	5900/14
Sharpless 264	(Lambda Orionis)	195.0	− 12.0	1400/25
NGC 2264	(Cone Nebula)	202.9	+ 2.2	2200/16
NGC 2244	(Rosette Nebula)	206.3	− 1.9	4700/110
NGC 2024	(Flame Nebula)	206.5	− 16.3	1300/12
IC 434	(Horsehead Neb.)	206.9	− 16.7	1300/25
Sharpless 2-276	(Barnard's Loop)	207.8	− 13.2	1300/350
NGC 1976/M42	(Orion Nebula)	209.1	− 19.4	1300/25
NGC 2327/IC 2177	(Seagull Nebula)	224.4	− 2.4	3800/170

- In Auriga, multiple FIR sources join the Tadpole and Fly Nebulae near (G173+1).
- In Orion, the HII region NGC 2174 (the Monkey Head Nebula) at (G190+1) shines brightly at FIR wavelengths. It appears to trace the right-most part of a large ring or shell of FIR sources centered at (G197+5) and spanning 13°.
- In Orion, the Lambda Orionis Nebula at (G195-12) sports an FIR-emitting shell of swept-up dust that surrounds the ionized hydrogen emission. Both are consequences of the extreme UV emission from the two hot O-type stars in the center.
- Again, in Orion, the Horsehead Nebula at (G201-16.7) and the Orion Nebula itself at (G209-19) appear to demarcate the lower right-hand edge of an even larger shell centered at (G212-12) that spans 15°.
- In Monoceros, the Rosette Nebula at (G206-1.9) marks the spot where the two candidate shells centered at (G197+5) and (G212-12), respectively, appear to meet. This coincidence in projection could be an optical illusion, however.
- In Canis Major at (G224-2), the HII region NGC 2327/IC 2177 (the Seagull Nebula) shows prominently in the FIR, likely due to the extensive dust content in this region.
- Three FIR arcs at positive latitude between 205° and 225° longitude appear illuminated by an external source at lower latitude and greater longitude. Closer examination of the present image, however, shows that the arcs break-up into many small sources of FIR emission, and so are likely energized internally.

Looking Opposite the Sun's Orbital Motion

The next and final quadrant of the Milky Way (Longitude = 220–320°) is visible mostly from the Earth's southern hemisphere (see Fig. 3.8). What better reason to travel to Chile, South Africa, Australia or New Zealand to catch these impressive views?! At optical wavelengths, dark clouds obscure much of the Vela and Crux constellations. The Coal Sack in Crux especially stands out against the background starlight. As shown in Fig. 3.9, the Quechua Indians of Peru made maps of the "dark constellations" that they could perceive, giving them the names of animals that were found in their environment. The Coal Sack was so named "the partridge." One reason it is so prominent is that it is relatively nearby, having a distance of only 590 light-years.

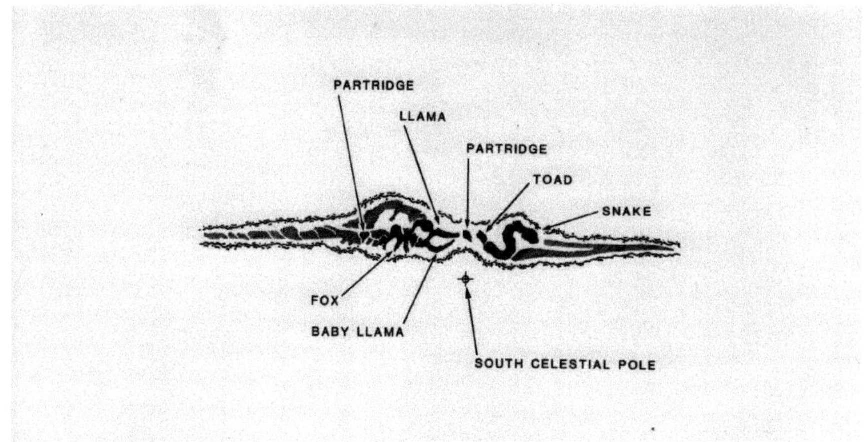

Fig. 3.9 "Dark constellations" along the Milky Way as delineated by the Quechua Indians of the Peruvian Andes (*credits* Drawing by J. Bieniaz [Griffith Observatory]; courtesy of E. Krupp [Griffith Observatory])

In the light of ionized hydrogen, the Gum Nebula sprawls across the Puppis and Vela constellations (see Fig. 3.10). It spans a whopping 35° of the sky, and at a distance of 1470 light-years, has a true extent of 900 light-years. It is thought to represent an "evolved" HII region complex that contains the Vela Supernova Remnant and perhaps other detritus from prior massive star-forming activity.

By far, the most dazzling region of emitting ionized gas in this quadrant is the Carina Nebula complex at (G288-0.8). Catalogued as NGC 3372, this star-bursting galactic ecosystem spans 2° (the angular equivalent of 4 full moons). The HII region's angular girth, at an estimated distance of 9000 ly, translates to an actual size of 300 ly. This powerful region could engulf the Orion Nebula 13 times over! It plays host to a panoply of marvels, including several newborn star clusters that contain hot ionizing stars, the Keyhole Nebula, and the Homunculus Nebula which itself surrounds Eta Carinae—an unstable supergiant star fated to explode as a supernova (see Fig. 3.11). We will revisit this fascinating region in Chap. 8 and Appendix B, where other starbursting galactic ecosystems are explored.

Along with the Gum and Carina Nebulae, this southernmost quadrant of the Milky Way includes other visibly prominent HII regions, as listed in Table 3.4.

Aside from the NGC, IC, and M designations for the optically visible HII regions, the Sharpless, Gum, and RCW catalogs have expanded the genre to about 500 such ionized nebulae in the Milky Way (see the caption to Fig. 3.10). Radio surveys have upped the tally to about 7000 emitting regions

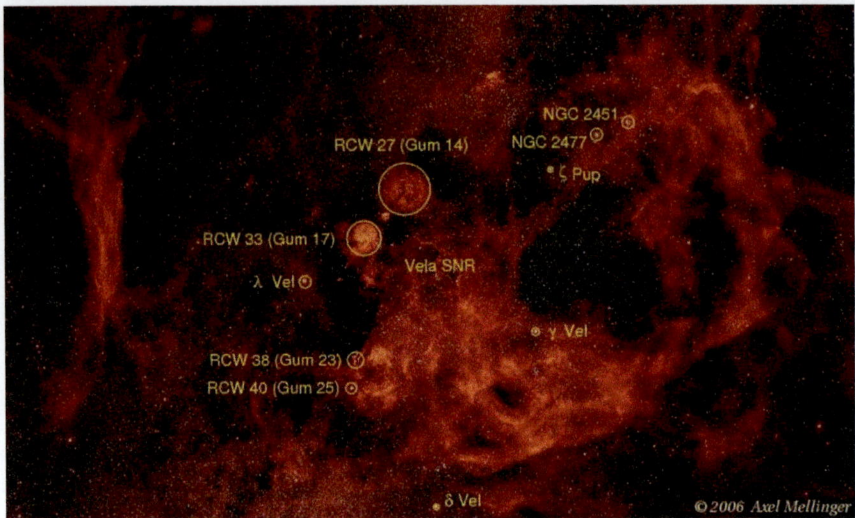

Fig. 3.10 The **Gum Nebula** in the light of ionized hydrogen. RCW refers to the catalogue of H-alpha emitting objects in the southern hemisphere that were compiled in 1960 by Alexander William Rodgers, Colin T. Campbell and John Bartlett Whiteoak from observations at Mount Stromlo Observatory in Australia. The RCW catalog is the southern complement to the Sharpless catalog that was compiled by Stuart Sharpless in 1959 from examination of images in the Palomar Sky Survey (*credit* Courtesy of Axel Mellinger, Central Michigan University, 2006)

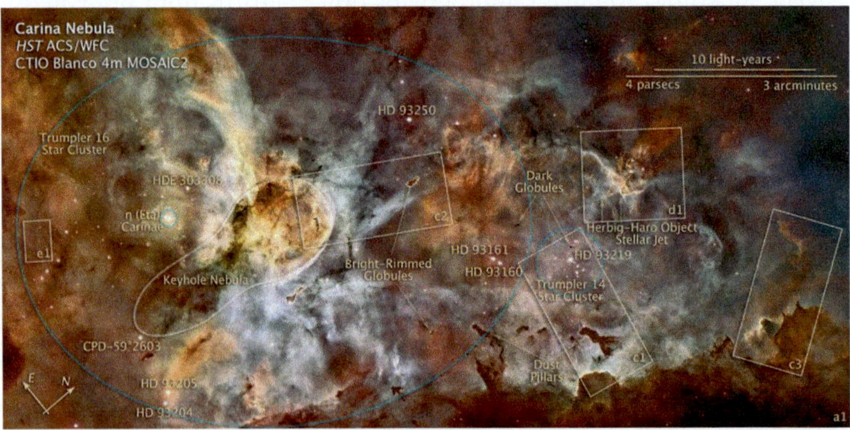

Fig. 3.11 A portion of the **Carina Nebula** complex showing some of its nebular and stellar denizens. This view combines multiple ground and Hubble observatory images in a 50-light-year wide panorama. Several of the labeled regions are featured with closeups in Fig. 6.4 and Appendix B.2 (*credits* NASA, ESA, Zolt Levay [STScI], Cerro Tololo Interamerican Observatory [NSF/NOIRLab])

Table 3.4 HII regions in the direction opposite the Sun's orbital motion (from Powell 2006)

Catalog name(s)	Common name	Longitude	Latitude	Distance/size (ly)
NGC 2327	(Seagull Nebula)	224.4	− 2.4	3800/170
NGC 2359	(Thor's Helmet)	227.6	− 0.3	20,000/60
NGC 2467		243.2	+ 0.4	15,000/35
NGC 2736	(Vela SNR [part])	266.9	+ 0.2	1500/9
NGC 3199		283.6	− 1.0	15,000/90
NGC 3372	(Carina Nebula)	287.5	− 0.9	9000/220
NGC 3576/3581		291.3	− 0.7	6000/50
NGC 3603		291.6	− 0.5	20,000/120
RCW 62	(Running Chicken)	295.0	− 1.7	6500/71

that are mostly obscured from visible view by foreground dust. The median normalized 100 μm images in Figs. 3.5, 3.6, 3.7 and 3.8 show these regions as well as star-forming nebulae with non-ionizing stars and the diffuse nebular aftermaths of all this activity. The FIR-emitting scene shown in Fig. 3.8 mirrors what was found in Fig. 3.6, where the view was in the direction towards the Sun's motion. Both views involve sampling galactic ecosystems that reside in the molecular ring as well as parts of the Perseus and Cygnus spiral arms beyond the ring. Highlights of this busy scene include

- A relatively evacuated region coinciding with the Gum Nebula that spans 20° and is punctuated by regions of secondary star formation along the cavity's rim.
- Lots of star-forming activity near (G268-1), some of which can be attributed to the HII region RCW38.
- A great void near (G270) that extends at least 30° in latitude.
- The Carina Nebula region becomes part of a much larger arc of star-forming regions that spans at least 8° in longitude.
- A possible "chimney" of enhanced diffuse FIR emission that lofts away from the Carina Nebula to positive latitudes.
- The Coal Sack at (G303+0) has no FIR emitting counterpart, perhaps because it is overwhelmed by more distant emitting regions.
- Multiple curved regions of diffuse FIR emission that may (or may not) indicate excavations by prior star-forming activity.

All this shaping of the diffuse emission can lead to one seeing physical structures that are not really there. For now, all we can conclude is that the Galaxy's dusty ISM is far from uniform, with irregularities that may or may not imply prior structuring by massive stars and the intense UV radiation, strong winds, and shocking explosions that they can inflict.

Putting It All Together

In prior parts of this chapter, I have alluded to many of the HII regions as residing within various spiral arms of the Milky Way galaxy. Unfortunately, no contemporary map of the Galaxy's disk can provide a reliable rendering of the HII regions and star-forming molecular clouds that have been detected to date. The challenge of ascertaining accurate distances to these nebular objects continues to frustrate the best of efforts. The HII regions that contain visible clusters of stars should enable distance determinations based on graphing the observed colors and brightnesses of the stars in the clusters. These so-called color-magnitude diagrams have strong diagnostic powers for determining both the distances and ages of the clusters (cf. Ch. 5 in Waller 2013). But even those HII regions with visible star clusters have distances compromised by the uncertain amounts of obscuring dust that dwells foreground and internal to the clusters. With these caveats, I show the following "God's Eye" maps to give you some idea of the context for all this star-forming foment within our home galaxy. Figure 3.12 shows a mapping of HII regions and clouds of neutral hydrogen that was compiled by Richard Powell as part of his online *Atlas of the Universe* (See http://www.atlasoftheuniverse.com/milkyway.html and references therein).

Names for the spiral arms are provided in Fig. 3.13, where significant artistic license has been exercised to render the overall morphology in the Galactic disk. Here, the four candidate spiral arms are seen to emanate from the ends of a central bar. Other mappings also invoke a central bar but favor a two-armed spiral pattern (cf. Plate 14 and Ch. 9 in Waller 2013).

We end this chapter knowing where to find and study active galactic ecosystems across the sky. We are less certain of their actual locations in 3 dimensions but remain optimistic that this situation will improve in coming years. In the next chapter, we will consider the nearest molecular clouds where the distances have been established to a fair degree of accuracy. Therein, we will behold cryogenic realms of nebular matter in remarkable tumult.

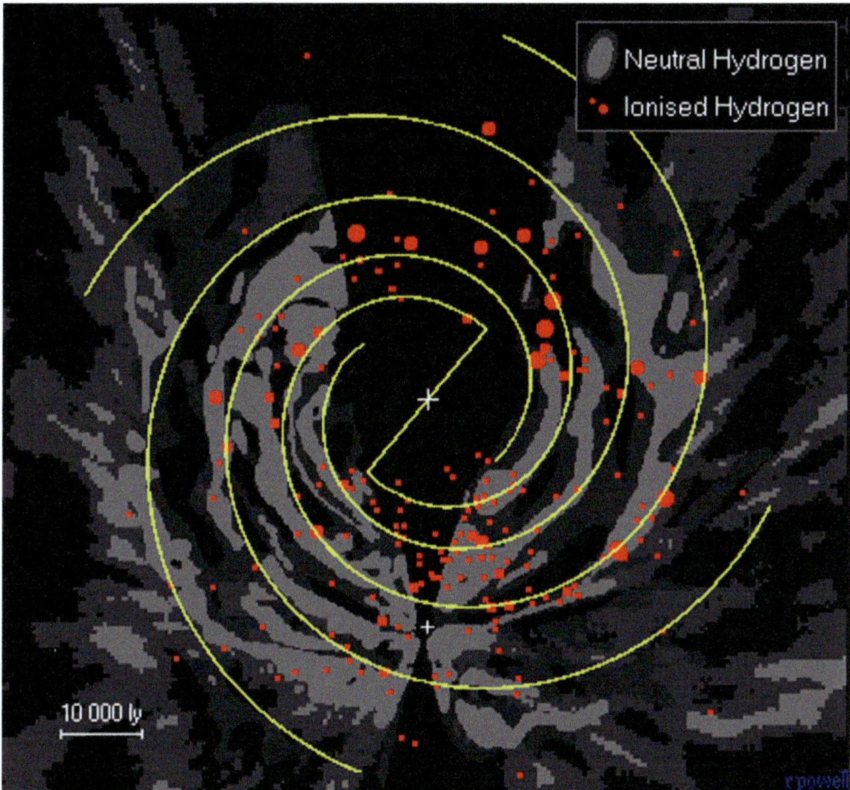

Fig. 3.12 A mapping of HII regions in the disk of our Milky Way galaxy with distances from the Sun based partly on the observed colors and brightnesses of stars in the embedded star clusters and partly on the HII regions' observed radial velocities referenced to an adopted rotation curve for the Galaxy (whereby the rotation speed is plotted as a function of distance from the galactic center). The HII regions are overlain upon a mapping of neutral hydrogen clouds, where all distances are based on the clouds' observed radial velocities referenced to the same rotation curve. The Sun is located at the smaller cross. Four candidate spiral arms have been drawn to guide the eye (*credit* R. Powell in his online *Atlas of the Universe* [http://www.atlasoftheuniverse.com/milkyway.html])

Fig. 3.13 Artistic rendering of structure in the Milky Way galaxy's disk based on the mapping of HII regions and neutral hydrogen clouds shown in Fig. 3.12 (*credit* R. Powell in his online *Atlas of the Universe* [http://www.atlasoftheuniverse.com/milkyway.html])

Part II

In Darkness Born—Molecular Clouds and Their Cosmic Progeny

How is it that the sky feeds the stars?

Titus Lucretius Caras in *On the Nature of Things* (55 BCE)

4

Content, Structure, and Dynamics of Molecular Clouds

We begin our more focused exploration of galactic ecosystems by examining those regions of darkness that can be seen in silhouette against the background of starlight in the Milky Way (see Figs. 3.1a and 3.5–3.8). By virtue of their silhouetting nature, these dark patches comprise the nearest of the known galactic ecosystems. They also tend to be of low mass relative to the larger more active regions that host bigger star clusters containing hot massive stars and associated HII regions.

A Bit of History

Up until the early 1900s, astronomers were uncertain about what they were seeing (or not seeing). Did the dark regions represent gaps in the stellar firmament, or were they physical clouds containing material that obscured the light from behind them? Resolution of this question began to emerge with Edward Emerson Barnard, a former parlor photographer who went on to become one of the most famous astronomers of his time. His magnum opus, *A Photographic Atlas of Selected Regions of the Milky Way*, involved taking photographs with a wide-field camera of his own design that he installed at Lick Observatory in California. He took special care to ensure that every photograph in his atlas was of the highest quality possible. Barnard died in 1923, four years before his great atlas was finally published.

I fondly recall the day in 1978 when I stumbled across one of these atlases in the reading room of the library at the Harvard-Smithsonian Center for Astrophysics in Cambridge, MA. I was immediately awestruck by the

W. H. Waller, *Crucibles of Creation*, Astronomers' Universe,
https://doi.org/10.1007/978-3-032-17258-7_4

exquisite beauty of the images in this atlas. For me, the dusty interstellar medium took on a visceral quality that has never left me. For Barnard, it took a while for him to conclude that the dark regions in his photographs were actual nebulae redolent with obscuring material. Remembering a partly cloudy night of observing in 1913, he noted that "I could not resist the impression that many of the black spots in the Milky Way are due to a cause similar to that of the small, black clouds mentioned above—that is, to more or less opaque masses between us and the Milky Way." Today, we can marvel at his atlas and subsequent identification of 349 so-called Barnard Objects (see Fig. 4.1).

By the mid twentieth century, Beverly T. Lynds of the National Radio Astronomy Observatory had taken it upon herself to update Barnard's listing of dark nebulae. She used photographs from the Palomar Sky Survey to make the identifications. Her subsequent "Catalogue of Dark Nebulae" contains 1802 listings, all visible from Mount Palomar. That gets you down to celestial latitudes of $-30°$ or so. Today, many astronomers still use her Lynds Dark Nebula (LDN) numbers to identify particular dark nebulae. For dark nebulae further south, they use the listing of 489 clouds by Feitzinger and Stuewe (1984) who examined photographic plates from the European Southern Observatory's Sky Atlas.

All this mapping at visible wavelengths served as an essential prologue for the mapping of emission from molecules resident in the dark nebulae. Molecular emission-line astronomy took off with the development of radio telescopes whose dishes were smooth enough to effectively focus emission at millimeter wavelengths, and whose detectors were sensitive enough to respond to the faint emission.

The workhorse molecule for the mapping was carbon monoxide (CO), a bright proxy for the far more abundant but mostly invisible diatomic hydrogen (H_2) molecules within the clouds.

By the 1980s, radio astronomers were completing large surveys of CO emission from molecular clouds in the Milky Way. By tracking any Doppler shifts in wavelength from the nominal 2.6 mm wavelength of the brightest CO emission line, astronomers could ascertain how the cloud was moving along the line of sight—and what sort of motions were occurring within the cloud. More focused examinations of particular clouds have been pursued in the light of other molecules such as carbon monosulfide (CS), hydrogen cyanide (HCN), formaldehyde (H_2CO), and ammonia (NH_3). Each of these sundry molecules trace different densities and temperatures within the clouds, thus enabling astronomers to diagnose the physical conditions inside. What they have found are dense nebular realms, as cold as they are dark. Densities

Fig. 4.1 **a** E. E. Barnard's photograph of the region surrounding the young star Rho Ophiuchi which appears amid nebulosity near the center of the field. The red supergiant star, Antares, shines to the lower left. Sinuous dark clouds, reflection nebulosity from illuminated dust, and a more distant globular cluster are all evident in this rich area of the Milky Way (*credit* E. E. Barnard 1927, https://exhibit-archive.library.gatech.edu/barnard/). **b** Corresponding chart of the Rho Ophiuchi region, as drawn by E. E. Barnard. Outlines of his catalogued dark nebulae are shown along with their B-numbers (*credit* E. E. Barnard 1927, https://exhibit-archive.library.gatech.edu/barnard/)

range from a few hundred hydrogen molecules per cubic centimeter overall to several million molecules/cc in the cloud cores. Typical temperatures are only a few tens of degrees above absolute zero. That is unless there is a hot newborn star nearby. Then the temperatures can skyrocket to hundreds and even thousands of degrees Kelvin.

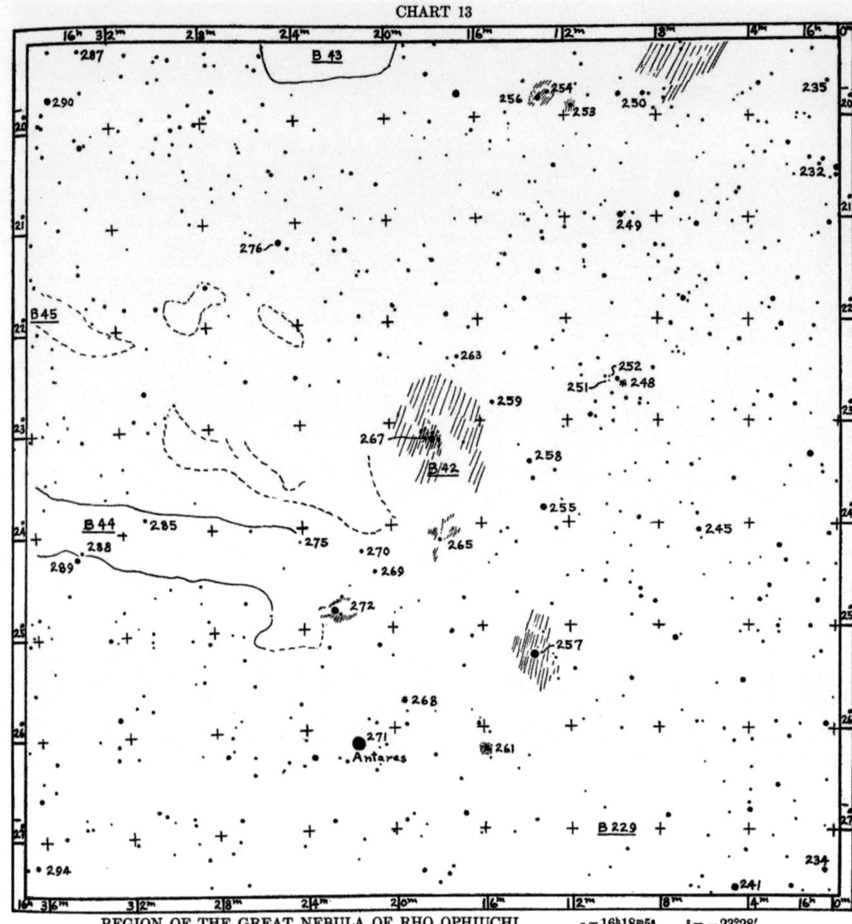

Fig. 4.1 (continued)

The Taurus Molecular Cloud and Local Bubble

An excellent case in point is provided by the Taurus Molecular Cloud. This relatively nearby and small cloud was first recognized as a region of dark nebulosity from Barnard's photographic survey (see Fig. 4.2).

Barnard's image reveals both patches and tendrils of dusty darkness. Detailed studies of the stars in this field have yielded estimates of visual extinction along the line of sight to each star. These estimates, in turn, have led to a mapping of the extinction produced by dust in the foreground nebulae (see Fig. 4.3). Whether a star is obscured or unobscured has also enabled astronomers to reckon distances to the obscuring matter, based on

Fig. 4.2 The **Taurus Molecular Cloud** from E. E. Barnard's *A Photographic Atlas of Selected Regions of the Milky Way* (*credit* E. E. Barnard 1927, https://exhibit-archive.library.gatech.edu/barnard/)

the distances to the most remote unobscured stars.. For most of the TMC, the resulting distance is a "mere" 430 ly.

The filamentary character of this mapping is further echoed with the mapping of emission from carbon monoxide molecules, as shown in Fig. 4.4. Here, measurements of Doppler shifts in the line emission reveal the cloud's line-of-site motions. These turn out to span about 12 km/s, significantly greater than those inferred from the cryogenic temperatures that have been measured in the molecular gas. Something other than thermal energy is motivating the nebular material towards supersonic speeds. Most astronomers look to interstellar turbulence as the motivating agent, though magnetic fields and winds from young stars could be adding to the chaotic mix.

Overall, the Taurus Molecular Cloud spans about 60 ly and contains roughly 24,000 solar masses of molecular gas. Within the cloud, a few hundred protostars and young stellar objects (YSOs) have been identified, some hosting pre-planetary disks and outflows. They amount to only 150

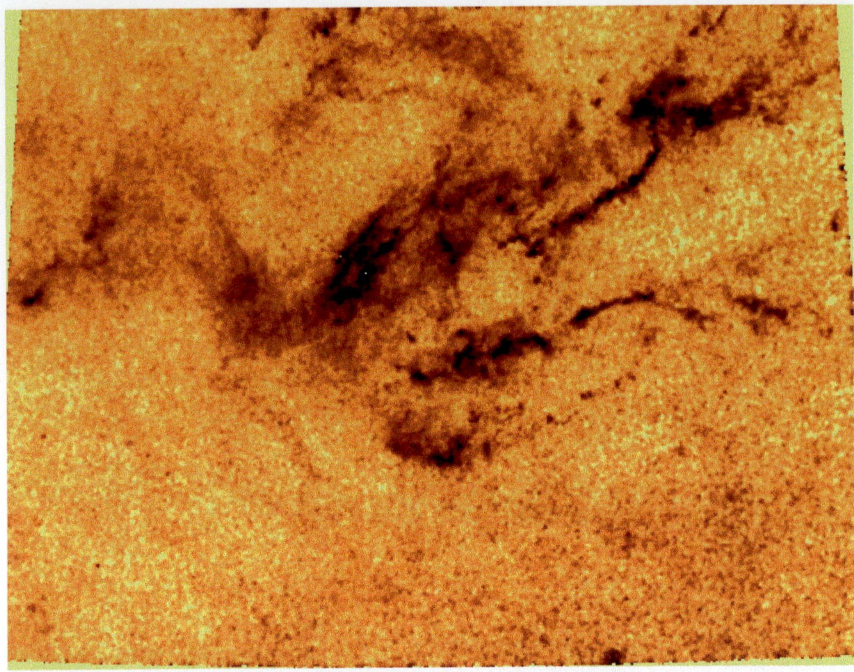

Fig. 4.3 Mapping of visual extinction to stars in the field of the Taurus Molecular Cloud. The dark patches and tendrils seen in Fig. 4.2 are even more evident, with many knot-like substructures apparent throughout the field (see also Fig. 2.1) (*credit* Courtesy of Paul Goldsmith [Jet Propulsion Lab, Caltech, NASA] with reference to Pineda et al. 2010)

solar masses or so, which suggests a low star-forming efficiency of 0.6% by mass. Clearly, this cloud is just getting started transforming itself into a cluster or association of young stars. In a few tens of millions of years, it will have dissipated into the Galactic aether.

Recent gauging of stellar distances obtained with data from the European *Gaia* space mission has shown that the Taurus Molecular Cloud forms part of a vast bubble which currently encompasses the Sun and Solar System. Most of the other nearby molecular clouds appear to populate the surface of the bubble (see Fig. 4.5). Within the bubble, hot diffuse gas has been confirmed from observations of faint but pervasive X-ray emission produced by the excited gas.

So, we appear to be bearing witness to some sort of powerful star-forming event having begun approximately 14 million years ago, whose dozen or so supernova explosions blew the bubble and incited further coalescence of molecular clouds and subsequent starbirth activity. The basic stats on these nearby molecular clouds are listed in Table 4.1.

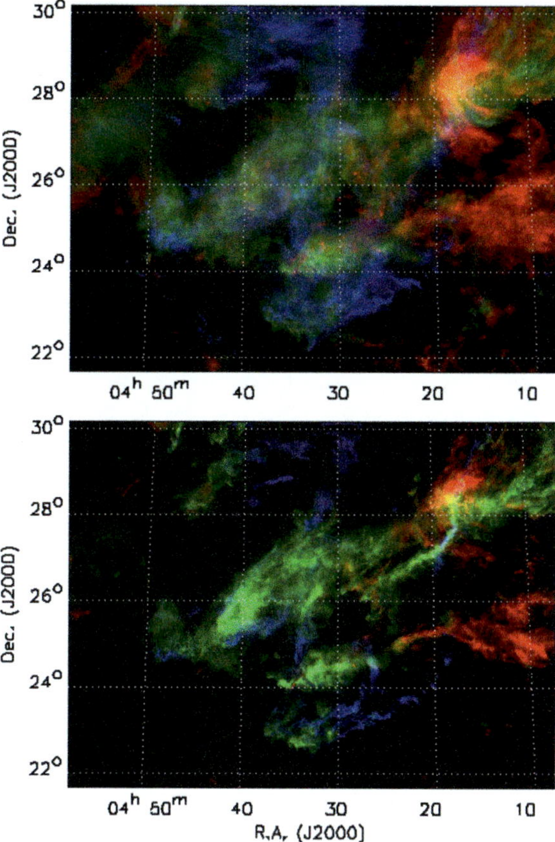

Fig. 4.4 Color-coded mapping of CO emission from the Taurus Molecular Cloud in the light of (top) ^{12}CO and (bottom) ^{13}CO. The top mapping from the most common isotopic species of CO best traces the diffuse molecular gas, while the bottom mapping is more sensitive to denser filaments, ridges, and blobs in the emitting gas. The color coding tracks the line-of-sight velocities, with blue colors indicating relative motions towards us and red colors indicating motions away from us. An overall gradient in radial velocities from left to right is suggestive of rotation, but more detailed investigations have not confirmed such a coherent pattern of motion (*credits* Paul Goldsmith [JPL/NASA], Chris Brunt [Exeter U.], Mark Heyer [UMass], Di Li [JPL/NASA], Gopal Narayanan [UMass], and Ron Snell [UMass]—the Five College Radio Astronomy Observatory, NSF)

These nebular neighbors will be revisited in Appendix A, where multiwavelength images of the clouds are presented and discussed.

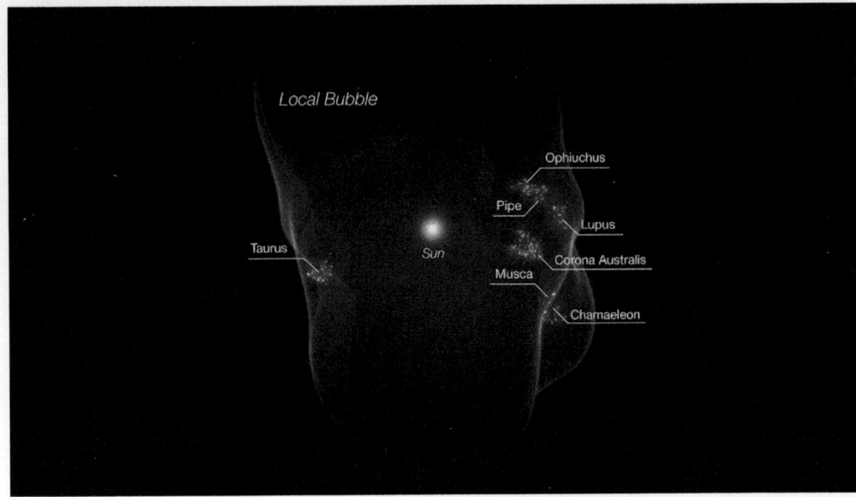

Fig. 4.5 Artistic rendering of the **Local Bubble** with associated molecular clouds populating its perimeter. The bubble spans approximately 1000 ly side to side. The Sun happens to be passing through this hot diffuse bubble at the current epoch (*credits* Leah Hustak [STScI], Catherine Zucker, Alyssa A. Goodman, Michael Foley, Douglas Finkbeiner (Harvard-Smithsonian Center for Astrophysics), and the Local Bubble Team [https://sites.google.com/cfa.harvard.edu/local-bubble-star-formation/home?authuser=0])

Table 4.1 Properties of molecular clouds within 600 ly of the Sun

Cloud name	Longitude (deg)	Latitude (deg)	Distance (ly)	Size (ly)
Taurus	169.8	– 16.0	427	60
Chameleon	297.1	– 15.5	610	32
Lupus	338.9	+ 15.8	498	61
Ophiuchus	353.1	+ 16.8	457	43
Pipe	357.1	+ 7.0	542	71
Corona Australis	359.8	– 17.7	492	43

Note The positional and distance data were obtained from Zhang (2023)

Tumult Within

The structure and internal motions of molecular clouds have vexed astronomers since the first interstellar detections of molecular line emission in the 1970s. Nowadays, we know that the clouds do not resemble anything close to tractable spheroids. Instead, they manifest exquisite filamentary morphologies that are often nested on multiple scales. There appears to be a fractal hierarchy of structuring at work within the interstellar medium, just as the statistics of the FIR-emitting froth indicated in the prior chapter.

As an astrophysicist once opined at a conference I attended, "The insides of molecular clouds are full of outsides."

To explain the baroque nature of molecular clouds, some astronomers point to large-scale instabilities in the Galactic disk that then incite a turbulent cascade down to smaller scales. Others point to local disruptors such as gravitational instabilities, stellar winds from newborn stars, intense UV radiation from hot massive stars, and supernova explosions marking the deaths of these same stars. All of these dynamics could be operating within giant molecular clouds; but even in the relatively small clouds considered in this chapter, we could be witnessing the leitmotifs of interstellar turbulence, gravitational instabilities, and stellar winds.

Thanks to the molecular emission lines having specific wavelengths of emission, astronomers can track the observed deviations in wavelength and interpret them as Doppler shifts involving motions towards or away from the observer. Once these motions have been thoroughly mapped across a molecular cloud, it is possible to analyze them in terms of the cloud's systematic vs. more random kinematics. Besides some indications of bulk motions (see Fig. 4.4), astronomers have found a kinematic cascade—with the greatest variations in radial velocity occurring over the largest scales (see Fig. 4.6). This sort of behavior is the clarion call of turbulence within the cloud. When comparing small and large molecular clouds, a similar relationship between the cloud's size and overall spread in radial velocities is evident. Here, the slope of that relationship is consistent with the clouds being gravitationally bound. Were the slope any higher, the clouds would be dissipating, and if any lower, they would be collapsing. Somehow, they are being stabilized by just the right amounts of internal turbulence and winds.

What I have so far avoided in this chapter are the causes and consequences of magnetic fields within molecular clouds. Whenever a charged particle moves relative to some frame of reference, it will produce a magnetic field that encircles the particle's trajectory. We understand that atoms have magnetic fields that depend on the net rotational motion of the electrons that whirl around the atomic nuclei. That is why iron atoms can be magnetically aligned in bulk to produce common magnets. Similarly, we attribute the Earth's magnetic field to the rotational motions of charged particles within the liquid metal outer core. The swirling Milky Way has an average magnetic field that is about a million times weaker than that experienced on Earth's surface. But in regions of more concentrated interstellar matter, the magnetic fields can grow to much greater significance. Moreover, they can both orchestrate and respond to any ordered motions in the concentrated gas. Recent

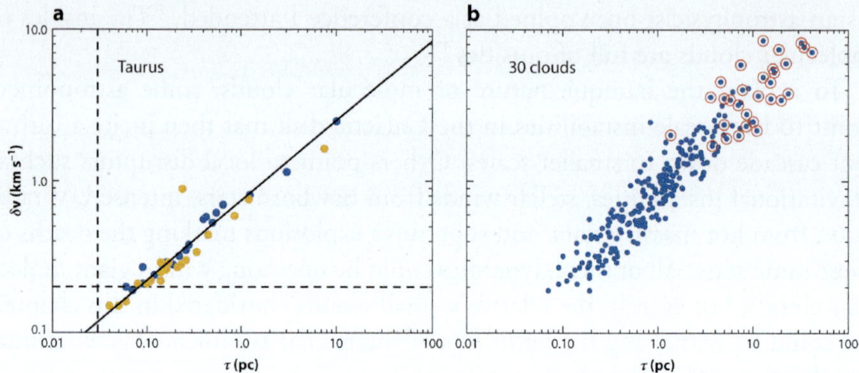

Fig. 4.6 Plotted relation between the spatial scale within a molecular cloud and the variation in radial velocity. The left-hand plotting includes only data from the Taurus Molecular Cloud. The right-hand plotting includes data from 30 clouds in the Solar Neighborhood as well as the Perseus spiral arm, thus expanding the range of spatial scales to greater values. The slope of this relation is consistent with a turbulent cascade operating within these molecular clouds (*credit* From M. Heyer and T. Dame 2015, Courtesy of Mark Heyer)

mappings of the magnetism in nearby molecular clouds have literally brought these dynamics to light (see Fig. 4.7).

You might wonder why cold molecular clouds would have charged molecules in them. It turns out that only a small fraction of the molecules

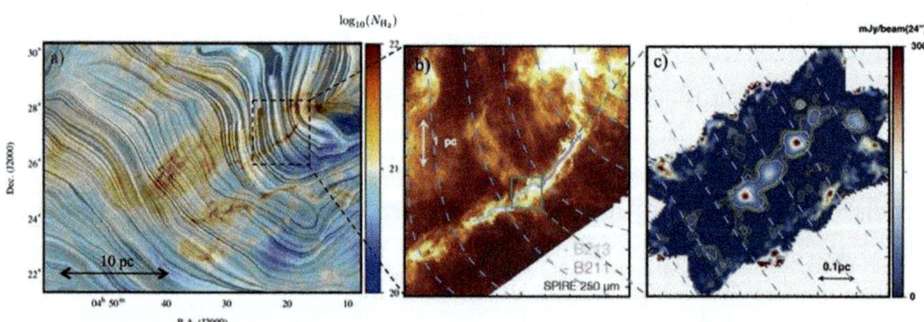

Fig. 4.7 Mapping of magnetic fields on various scales in the Taurus Molecular Cloud. The left-hand and middle maps show the dust emission at 250 μm wavelength obtained with the *Herschel* space observatory. Overlain are magnetic fields derived from the polarized emission from dust at a wavelength of 850 μm as observed with the *Planck* space observatory. The right-hand map shows a zoomed-in region of emitting cloud cores observed at 1.2 mm wavelength with the *IRAM* 30 m radio telescope. Overlain are magnetic fields inferred from the *Planck* mapping. In this part of the cloud, at least, the magnetic fields are oriented perpendicular to the filaments—thus indicating possible flows towards the filaments (*credit* From P. André et al. (2019) with permission)

are charged—by impacts from cosmic rays ripping through the Galaxy. But that is enough to sustain the magnetic fields that are observed. It is also sufficient for these fields to guide flows of molecular gas along them, as appears to be the case with the magnetized flows associated with filaments in the Taurus Molecular Cloud. But mappings of magnetic fields in other molecular clouds show differing configurations of the magnetic fields relative to the molecular features. We are left grappling with a more complicated situation of mutually interacting agents, whose detailed *magnetohydrodynamics* have yet to be adequately understood.

In the next chapter, we will revisit the fascinating behavior of magnetic fields but on much smaller scales associated with protostellar inflows. And in subsequent chapters involving much larger galactic ecosystems, we will again consider magnetic fields in the context of energetic outflows, ie. starburst winds.

5

Cloud Cores and Protostars

After pondering the puzzling filigree that characterizes molecular clouds, one might expect that closer examination would yield ever smaller filaments, sheets, and blobs nested within—like so many nested matryoshka dolls of ever diminishing size. But that is not the case. Instead, high-resolution mapping by the most powerful array of telescopes operating at submillimeter wavelengths has revealed a lower limit of structuring in the form of discrete cloud cores with substructures of only a few hundredths of a light-year in extent (see Fig. 5.1). This size scale is equivalent to a mere few thousand Astronomical Units, well within our own Solar System's Oort Cloud of comets. We are now at the threshold of beholding the transformation of molecular cloud cores into protostellar systems.

Until recently, astronomers had no direct way to investigate these molecular motes. Instead, they had to work with models that track the distillation of dense cores out of the nebular foment extant within molecular clouds. They were faced with a multiplicity of contending influences, each of which could play important roles in fostering or hindering the emergence of molecular cloud cores. Paramount among these agents, self-gravity will rule the roost in the absence of any other opposing forces. But that never happens, as a panoply of energetics are always active within the clouds. For small molecular clouds, astronomers need not concern themselves with the powerful effects of newborn massive stars. That leaves the thermal energy of the nebular gas, rotational motions, the ambient turbulence and associated shock heating within the cloud, the magnetic fields that thread through the cloud, and any winds from newborn low-mass stars.

W. H. Waller, *Crucibles of Creation*, Astronomers' Universe,
https://doi.org/10.1007/978-3-032-17258-7_5

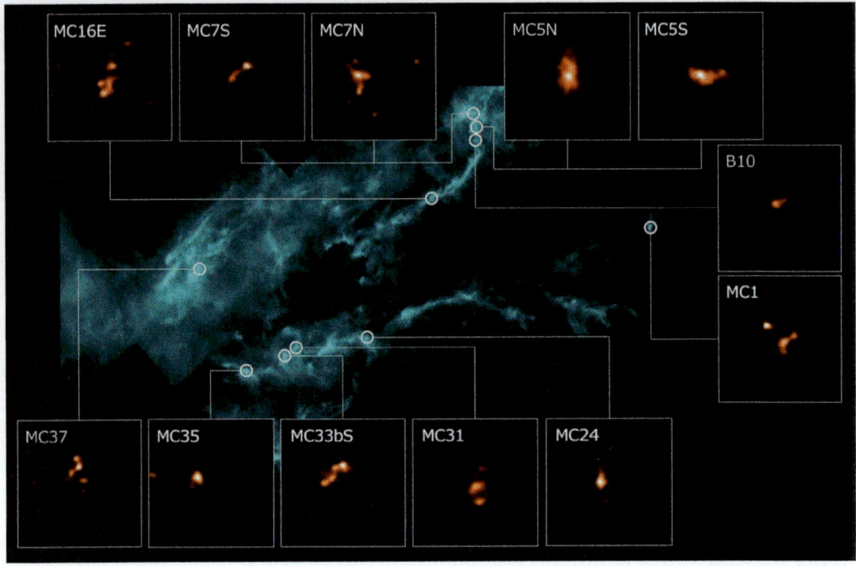

Fig. 5.1 The **Taurus Molecular Cloud** as imaged by the *Herschel* space observatory at 500 μm wavelength with closeups of cloud cores mapped by the *Atacama Large Millimeter/submillimeter Array (ALMA)* at 1 mm wavelength, what the investigators called "eggs". Several of the emitting cores sport substructures, each of which could host the development of a protostar (*credits ALMA* [ESO/NAOJ/NRAO], Tokuda et al., ESA/Herschel)

Aiding the inward directed force of self-gravity, radiation from the gas and dust can release energy and so help to reduce the temperature of the gas. All of these influences vie in ways that are both complex and confounding. What we do know is that many molecular clouds contain discrete cores and that most of these cores have yet to collapse in free fall. Given that the free-fall time for a dense core is estimated to be a mere 10,000–100,000 years, we might expect that most cores would have collapsed already. Instead, we can see that most cores appear propped up against collapse. To venture further, it is necessary for us to entertain some basic cloud physics, beginning with …

Changing Phase

Before there were molecular clouds laden with diatomic hydrogen (H_2), our Galaxy was awash with primordial atomic hydrogen (HI). Today, about half of the hydrogen gas in our part of the Milky Way is in the molecular phase. How then did half of the primordial atomic gas change its phase? Getting individual atoms of hydrogen to interact and bind together while in the gas phase is regarded as a very tall order. The low densities preclude most meet-ups, and the thermal energetics—while low—are still too high to encourage binding. Many astronomers look to dust in the nebular mix as the essential matchmakers. Atoms of hydrogen can more readily stick to a microscopic grain of dust and then hook-up with another atom on the same grain. Moreover, the atoms can release some of their thermal energy to the grain and so prime themselves for binding. Lastly, the dust protects the molecules from any destructive UV radiation in the region. The denser the cloud, the more efficient is the transformation from the atomic to molecular phases. That is why many models of molecular clouds invoke a molecular interior surrounded by an atomic halo (see Fig. 5.2).

Condensing

Much of the molecular cloud's subsequent structure can be explained as a condensation process, whereby gravity amplifies any initial over- and under-densities. The clumps get clumpier, while the voids get emptier. Winds from the first stars to form, as well as internal turbulence, and magnetic fields can mediate this consolidating affair. Soon, the cloud resembles a baroque assemblage of blobs, filaments, and sheets. We are now ready to witness the distillation of cloud cores and the subsequent birth of stars.

Fragmenting

Once gravity gets the upper hand within any condensed parcel, that region is subject to fragmentation. Sometimes, the geometry of the parcel can dictate the scale of fragmentation. For example, the width of a filament (about 0.3 ly) can provide the upper limit for the fragmenting scale. Cloud cores with spacings consistent with that scale would then develop, dotting the filament like beads on a string (see Figs. 2.1 and 4.6). In the absence of any geometric directive, the scale of fragmentation can be surmised from the physics of gravitational instabilities (see Appendix E.1). From these considerations, one can see that the minimum radius for gravitational collapse increases as the square root of the temperature (T) and decreases as the square root of the particle density (n), ie. as $(T/n)^{1/2}$. The corresponding minimum mass for gravitational collapse increases as the temperature to the 3/2 power and decreases (again) as the square root of the particle density, ie. as $T^{3/2}/n^{1/2}$. In other

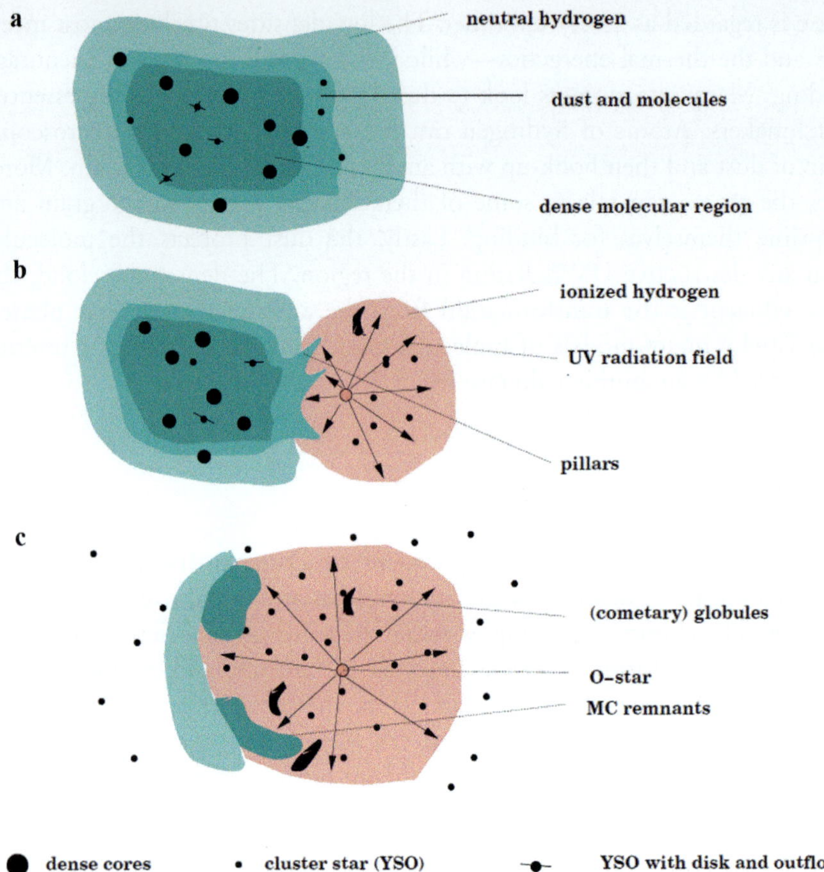

Fig. 5.2 Idealized schematic of an evolving molecular cloud—with its atomic halo, molecular interior, and dense cloud cores. Here, the cloud is big enough to host the formation of a massive O-type star whose intense UV radiation field photo-ionizes a portion of the cloud while photo-excavating cavities and pillars in the exposed "surface" of the cloud (*credit* From N. S. Schulz (2006) *"From Dust to Stars ..."* with permission)

words, cool and dense clouds will be more prone to fragmenting into small low-mass parcels that are gravitationally unstable to collapsing.

Figure 5.3 shows the minimum radii and masses for gravitational collapse as a function of particle density and temperature. These plots indicate that densities greater than a few thousand particles/cm^3 and temperatures less than 30 K are more favorable to generating cloud cores amenable to collapsing into one or more protostars.

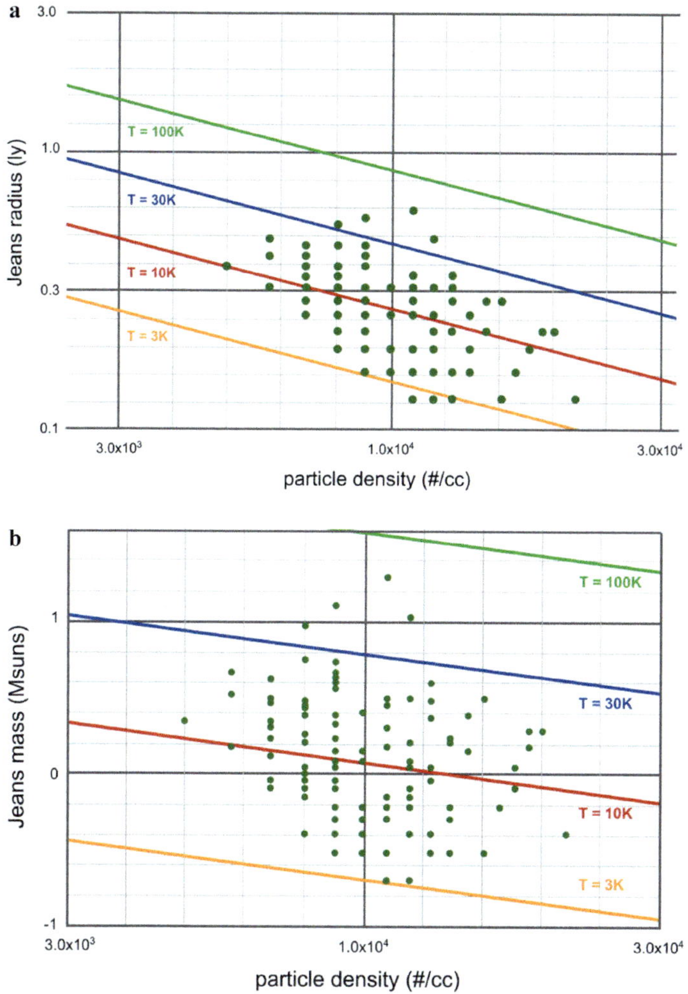

Fig. 5.3 **a** Plot of the minimum radius for gravitational collapse (Jeans radius) as a function of the H_2 number density for cloud temperatures of 3, 10, 30, and 100 K. Here, the plotting is in a log–log format, so that the $(T/n)^{1/2}$ dependence results in parallel straight lines (See Appendix E.1 for full derivation). The particle densities and radii of actual cores in the Pipe Nebula are overplotted for comparison. Given a typical core temperature of 10 K, many of the cores are seen to have radii exceeding the Jeans radius and thus should be unstable to gravitationally collapsing in the absence of other agents at play. **b** Plot of the minimum mass for gravitational collapse (Jeans mass) as a function of the H_2 number density for cloud temperatures of 3, 10, 30, and 100 K. Again the plotting is in a log–log format, so that the $(T^{3/2}/n^{1/2})$ dependence results in parallel straight lines (see Appendix E.1 for derivation). The number densities and masses of actual cores in the Pipe Nebula are overplotted for comparison. Given a typical core temperature of 10 K, many of the cores are seen to have masses exceeding the Jeans mass and thus should be unstable to gravitationally collapsing in the absence of other agents at play (*credit* W. H. Waller and L. H. Slingluff, with data from J. M. Rathborne et al. 2009)

Collapsing and Contracting

The minimum time for a cloud core to collapse is that resulting from free-fall. This timescale depends inversely on the square root of the average density, with typical cloud cores having free-fall timescales of only 10,000–100,000 years (see Appendix E.2). Such short timescales would suggest rampant collapsing into stars with few cloud cores remaining. The fact that we can directly observe plenty of cloud cores inside molecular clouds contradicts this suggestion and so requires other mechanisms to slow the collapsing process.

Another issue with gravitational instabilities and the fragmenting that ensues is that further condensation will lead to higher densities which, in turn, will cause further fragmentation into ever lower mass parcels. This runaway scenario would play out were it not for the cloud cores' emerging autonomy. Once the cores get sufficiently dense, they develop a variety of mechanisms for resisting further condensation and fragmentation. First, the cores become opaque to their own thermal radiation. As the dust in the cores gets denser, it effectively traps the radiation and heats the core, thus providing an internal pressure against rapid gravitational collapse. Instead, the cloud core begins to contract at a much slower rate. An equilibrium of sorts then plays out as the escaping radiation tracks with the slow release of gravitational potential energy during the contraction. This so-called Kelvin–Helmholtz contraction can take upwards of 30 million years for a Sun-like protostellar core, thereby nixing any further fragmentation and explaining why we can observe these cores in the first place (see Appendix E.3). More massive cloud cores contract far more quickly, as is shown in Fig. 5.4. Indeed, the highest-mass cores and protostars have such fleeting lives as to be rarely ever detected.

Other prophylactic agents against rampant collapse include the magnetic fields that get intensified during the condensation phase. They provide an additional pressure against gravitational infall, but only in directions perpendicular to the field lines. This results in a prevalent direction of inflow that trends parallel to the field lines, as appears to be the case shown in Fig. 5.5. Eventually, the magnetic fields leak out and dissipate via a process known as ambipolar diffusion. Over a period of 1–10 million years, the charged and neutral particles interchange energies in such a way for the neutrals to slowly drift inward past the magnetically fixed ions. Finally, the kinetics of rotation, turbulence, and winds from previously formed stars can slow things down further. We now can explore in a more leisurely way the various stages of

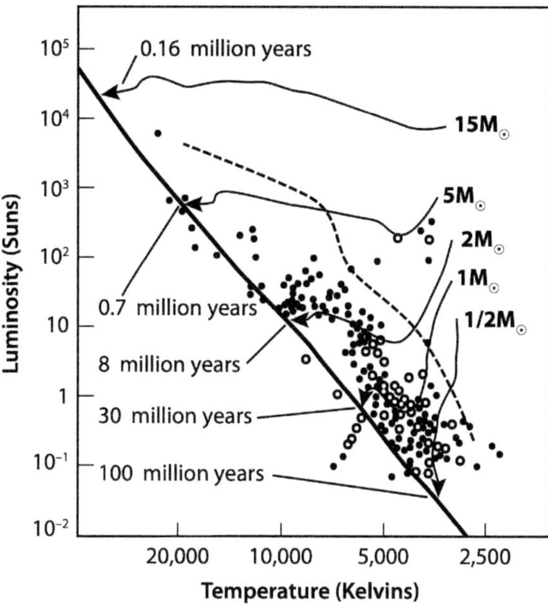

Fig. 5.4 Luminosity versus temperature (Hertzsprung-Russell) diagram of the young star cluster NGC 2264 showing the evolutionary tracks of protostars and pre-main-sequence stars. The nearly straight diagonal line is the Main Sequence, where fusion of hydrogen powers most of the stars familiar to us. The higher luminosities of the young stellar objects (YSOs) results from powering by the conversion of gravitational energy into radiation during the contraction phase. The dashed "birthline" delineates the evolutionary transition from a protostar to a pre-main-sequence star for differing stellar masses. High-mass stars evolve to the Main Sequence hundreds of times faster than Sun-like stars (*credits* From W. H. Waller (2013), p. 137 with reference to *Horizons*, 7th edition by M. Seeds, Pacific Grove, CA: Brooks/Cole Publishers (2002), pp. 162, 164)

protostellar growth leading to pre-main-sequence stars that are girdled with pre-planetary disks.

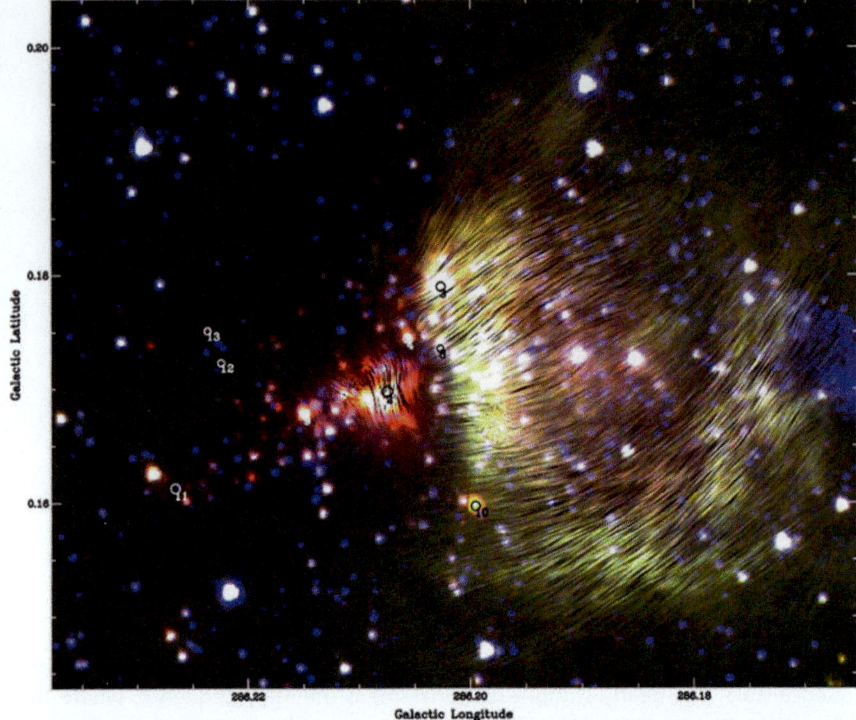

Fig. 5.5 The star-forming cloud BYF 73 as imaged in the mid-infrared by the *Spitzer Space Telescope* with magnetic field lines from *SOFIA*/HAWC+ overlaid. Circled regions are protostars identified with the *ALMA* and *Gemini* observatories. This stellar nursery manifests one of the strongest inflows known, mostly toward the monster protostar MIR 2 (*credits Spitzer Space Telescope, SOFIA,* NASA/*Spitzer*/*SOFIA*/*ALMA*/*Gemini*/*AAT*/Barnes et al.)

6

Young Stellar Objects and Proto-Planetary Systems

Once a dense molecular cloud core develops a protostar, the stage is set for further evolutionary marvels. Initially, the protostar cannot be detected visually, as it is completely swaddled within its nebular birth cloud. At temperatures of only 10–100 K, it emits mostly at sub-millimeter and far-infrared wavelengths. Astronomers have situated these infant stars at the beginning of an evolutionary sequence of Young Stellar Objects (YSOs) which they have divided up into four classes based on their spectroscopic properties (see Fig. 6.1).

Class 0 YSOs mark the threshold of protostars gestating within the molecular cloud cores. They are in hydrostatic equilibrium, such that they are neither collapsing nor expanding significantly. But there are lots of other processes affecting them. First, they are slowly contracting, and in doing so, emitting gravitational energy in the form of sub-mm and far-IR emission. The corresponding temperatures of the emitting matter have risen to 20–70 K. Second, the enveloping nebulosity continues to flow inward and accrete onto the protostar. The geometry of this accretion depends on the protostar's degree of rotation. Because angular momentum must be conserved within the pre-stellar cloud core, any initial rotation will be amplified as the matter concentrates to a smaller radius. The same phenomenon occurs whenever a spinning skater brings in her arms or a diver tucks into a crouch as part of his somersault. Sometimes a binary protostellar system arises from all the condensing and collapsing. Here, the orbital motion of the two protostars takes on much of the system's angular momentum, thereby allowing the individual protostars to continue contracting without spinning up too radically.

W. H. Waller, *Crucibles of Creation*, Astronomers' Universe,
https://doi.org/10.1007/978-3-032-17258-7_6

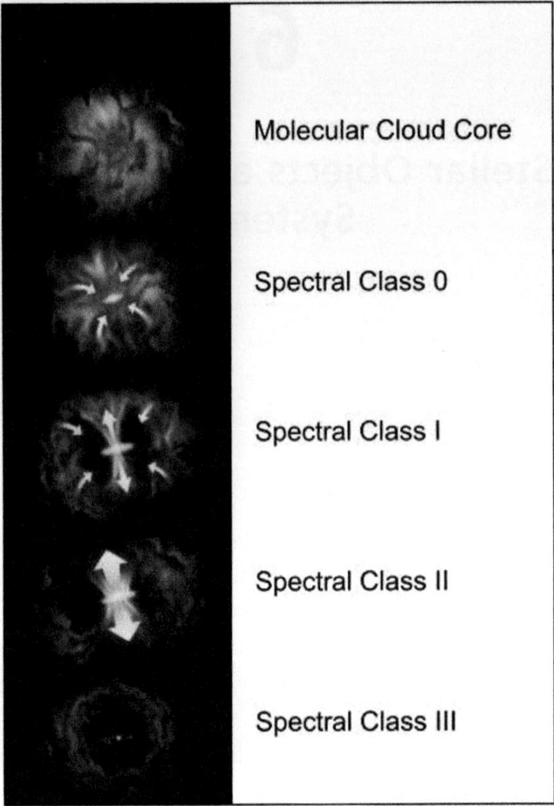

Fig. 6.1 Stages of stellar gestation (from top to bottom) beginning with a molecular cloud core, the infall of gas to make a protostar and accretion disk (Class 0), the emergence of a bipolar outflow from the protostar (Class I), the resultant clearing of material in the protoplanetary disk (Class II), and the residual star with orbiting planets that characterize the Solar System and other "mature" exoplanetary systems (Class III) (*credits* Courtesy of Charles Lada [Harvard-Smithsonian Center for Astrophysics] and Rob Wood [Illustrator])

The remnant magnetic field may also play an important role in guiding the accretion of matter. These mediating agents may constrain any inflow along the rotational axis and/or magnetic field lines. The upshot is the development of an asymmetric protostellar mass that will next concentrate into a more spherical protostar surrounded by an accretion disk. All this congealing lasts about 30,000 years for a Sun-like protostellar system.

Class I YSOs denote those protostellar systems with recognizable components. Were we to zoom inward through the molecular cloud core, we would see that the distended protostellar system has gravitationally settled into an even denser spheroid with a surrounding disk of accreting gas and dust (see

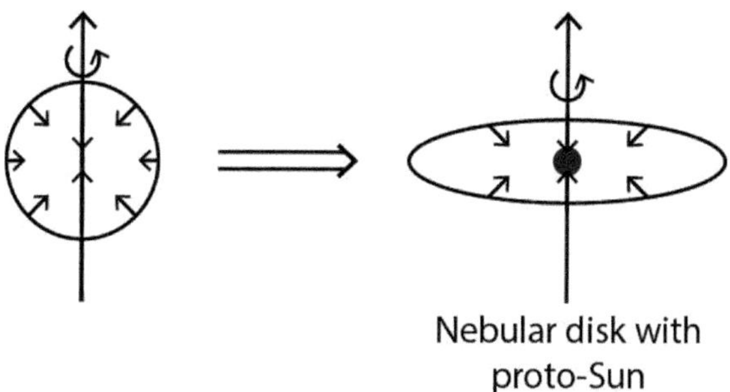

Fig. 6.2 Cartoon of a rotating cloud core that collapses from the inside out via its own self-gravity into a central mass and enveloping disk, as first articulated by Pierre-Simon Laplace in the early 19th Century. The cloud core's self-gravity must bind its rotation which is greatest at the equator. This binding force reduces the gravitational force available for infall at the core's equator, thus producing a flattened disk. Further collisions between the gas particles in the disk will dissipate the vertical kinetic energies, resulting in a thinner disk in circular rotation. The central mass will become a self-luminous star, while the thin disk will ultimately break-up into several planets (*credits* W. H. Waller and L. H. Slingluff)

Figs. 6.1 and 6.2). As the matter has concentrated, the temperatures have ramped up to values of 70–650 K. The resulting spectrum reaches its peak emissivity at far-infrared wavelengths, but also has a noticeable contribution at shorter near-infrared wavelengths from the hotter embedded protostar itself. Matter from the natal cloud core continues to flow into the accreting system, while outflows begin to exert their influence on their surroundings. These outflows can carry away a significant fraction of the protostar's initial angular momentum, again helping the protostar to avoid spinning-up to untenable velocities. A Sun-like protostellar system could undergo this intense mix of inflows and outflows for around 200,000 years.

Class II YSOs manifest the transition from protostellar to pre-main sequence stellar systems. Enough of the birth cloud has been blown away to expose the central star to visible view. Now, the spectrum is dominated by light at visible and near-infrared wavelengths that correspond to temperatures approaching 3000 K. A pre-main-sequence star of near-solar mass would be called a T Tauri star, named after the prototype that was first observed to be strangely variable in 1852. Often, a longer wavelength tail in the spectrum betrays the presence of a protoplanetary disk at lower temperatures of around 1000 K or less. Outflows along the rotational axis grow especially pronounced, resulting in ionized jets of material impacting their surroundings and shock heating them. The resulting *Herbig-Haro objects* have been a hallmark of protostellar and pre-main-sequence stellar evolution since they were first recognized by

Fig. 6.3 **NGC 1333** as seen at mid-infrared wavelengths. This galactic ecosystem resides a thousand light-years away in the constellation of Perseus. Besides hosting scores of pre-main-sequence stars, NGC 1333 sports many jets that have been identified as Herbig-Haro (HH) objects. These shocking bipolar outflows are coded green in this image from the Spitzer Space Telescope (*credits* NASA/JPL-Caltech/R. A. Gutermuth [Harvard-Smithsonian CfA])

George Herbig and Guillermo Haro in the 1940s (see Figs. 6.3 and 6.4). Astronomers estimate that this T-Tauri stage could last upwards of a million years for stars of solar mass.

Class III YSOs represent the last recognizable stage of the young stellar genre. By now, the protoplanetary disk has gravitationally congealed into planetesimals that will eventually amalgamate into planets. A remnant debris disk of

Fig. 6.4 A rogue's gallery of Herbig-Haro objects. (Top left) **HH 24** hosts two collimated jets shooting out from an obscured protostar (NASA/ESA/HST, D. Padgett [GSFC], T. Megeath [U. Toledo], and B. Reipurth [U. Hawaii]). (Top right) **HH objects 901 and 902** in the Carina Nebula (NASA/ESA/HST, Mario Livio [STScI]), (Bottom left) **HH 211** in the near-infrared light of shocked molecular hydrogen, carbon dioxide, and other molecules (NASA/JWST, T. Ray [Dublin]), (Bottom right) **HH 212** includes luminous knots from episodic outbursts by the hidden protostar [ESO/M. McCaughrean])

rocky and icy shards patrol the outer limits of the system. The Kuiper Belt just beyond the orbit of Neptune in our own Solar System might represent just such a remnant debris disk. The spectrum of these YSOs is dominated by the pre-main-sequence (PMS) star with very little excess at infrared wavelengths. This PMS star can still produce dramatic outbursts, as its magnetic activity has yet to subside. Indeed, one of the best ways to identify PMS stars is via the X-ray emission that they produce.

From the Mouths of Babes

It is worth our while to pause a bit and ponder the incredible outflows being jettisoned by these young stellar objects. The bipolar Herbig-Haro (HH) objects trace the business ends of powerful jets slamming into their natal surroundings. Clocked at several hundred km/sec, the jets shock heat the ambient nebulosity eliciting fluorescent alarms in the light of ionized hydrogen, sulfur, and oxygen at optical wavelengths and of iron, molecular hydrogen and other molecules at near-infrared wavelengths (see Figs. 6.3, 6.4 and 6.5). Collimated by intense magnetic fields at their protostellar source, the jets provide an important means for carrying away angular momentum and thus permitting the protostars to continue contracting into PMS stars and ultimately main-sequence stars. As shown in Fig. 6.5, the ionized and molecular outflows extend well beyond the gestating protostar, thus impacting much of the hosting cloud. Given enough of these outflows, the cloud can be irreparably shredded into a mess of filaments and sheets.

Disk Worlds

The first evidence for protoplanetary disks girdling young stellar objects came in the form of spectra that featured near-infrared excesses. At the time, it was not clear what was absorbing the optical light from the YSOs and re-radiating the energy in the infrared. Circumstellar disks would qualify, but enshrouding clouds could also produce the observed IR excesses. It took high-resolution imaging to verify the presence of dusty disks around the gestating protostars.

The first images to reveal something "disky" surrounding the YSOs came from ground-based telescopes beginning in 1979. Images of the Orion Nebula taken with the 1 m telescope at Pic du Midi Observatory in France revealed line-emitting "condensations" that the investigators interpreted as "globules" that were being externally ionized (Laques and Vidal 1979). It

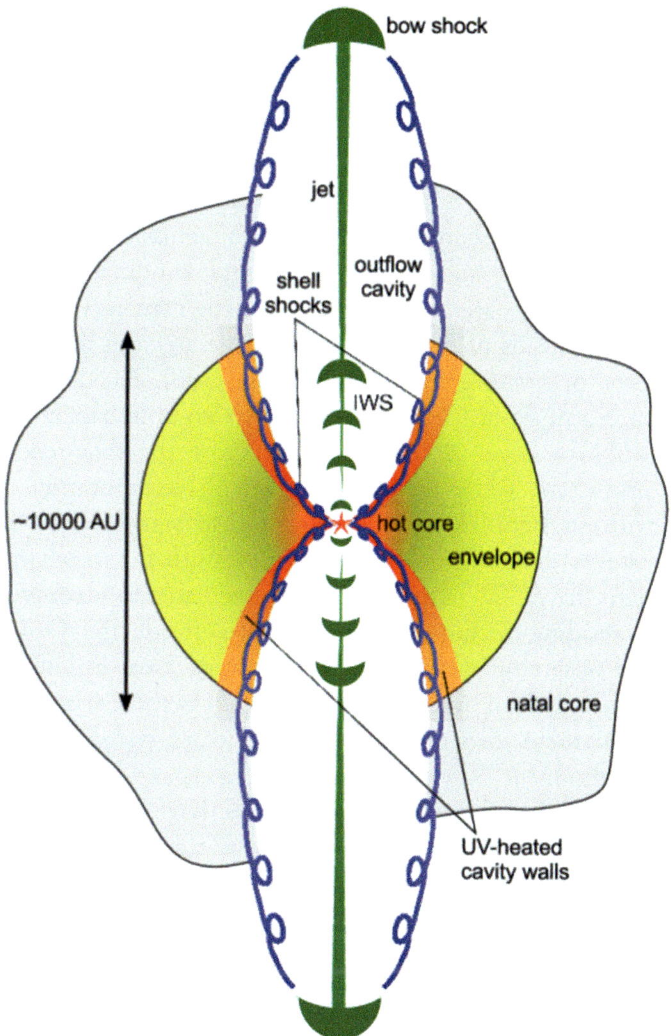

Fig. 6.5 Cartoon illustration of a bipolar outflow originating from within a protostellar molecular core, with the different physical components and their nomenclature indicated. IWS stands for internal working surfaces. The arc-like bow shocks along the jets trace episodes of outbursts from the YSO. The indicated scale is appropriate for a low-mass YSO. On this scale, the ~100 AU radius disk surrounding the protostar is not visible. More detailed studies have revealed correspondingly greater complexity in the outflows, as they are both spatially and temporally variable on multiple scales. On small spatial scales the narrow jets and knots can vary on timescales of weeks to years, while on much larger spatial scales the molecular outflows can vary over decades to centuries (*credit* Adapted with changes from E. F. van Dishoeck et al. 2011)

took the much greater acuity of the refurbished *Hubble Space Telescope* (beginning in 1995) to resolve these sources into their marvelous components (see Fig. 6.6). As shown in the *HST* images, cometary nebulae (resembling the "poor unfortunate souls" in *The Little Mermaid* movie) betray the intense erosive processes that the *Trapezium* of central hot stars is inflicting upon them (see Fig. 6.6a). The hottest and most massive star, Theta 1 Orionis C, leads the disruptive charge. Its blazing UV emission is thought to be producing the cometary tails, while its mighty winds are thought to be creating the bow shocks seen near the heads of the cometary nebulae.

In closeup, the heads of the cometary nebulae are seen to contain protoplanetary disks, or *proplyds* for short (see Fig. 6.6b). Backlit by the Orion Nebula's fluorescent gases, these disks manifest an abundance of obscuring dust. The disks in edge-on configurations show the dust-reddened light of central protostars, while the disks in face-on configurations reveal the protostars without significant reddening—a clear sign that we are dealing with circumstellar disks of dusty composition. With the installation of new cameras aboard the *HST*, astronomers have been able to identify more than 40 proplyds residing within cometary nebulae inside the Orion Nebula. Using similar observing techniques, astronomers have found comparable collections of proplyds in other HII regions, such as the Carina Nebula, thus confirming that these sorts of dusty protoplanetary systems are rife within star-forming regions.

Even finer details of protoplanetary disks have come from *ALMA*'s mapping of the dust and molecular emission at submillimeter wavelengths. Here, the disks resolve into myriad structures—including concentric rings, spirals, and smoothly varying "frisbees." (see Figs. 6.7 and 6.8). For example, the disk in the relatively nearby HL Tauri protostellar system displays rings whose gaps are thought to indicate the presence of protoplanets that have cleared out their respective annuli.

To go any deeper into the disks and their structuring, astronomers currently rely on models that take into account the presence of gas, dust, planetesimals of various sizes, and full-fledged planets—all in orbit around their host star and feeling its radiative effects to varying degrees. Magnetic fields further complicate their models, but to first order, astronomers have settled on an architecture that features a heterogenous disk whose thickness flares at greater radii (see Fig. 6.9). Dust and rocky planetesimals prevail close to the star, as they can withstand the star's heating effects. Farther from the star, ices of various sorts add to the mix. Beginning with the snowline where water begins to freeze out, snowlines of ammonia, methane, carbon dioxide, and carbon monoxide make their sequential appearances.

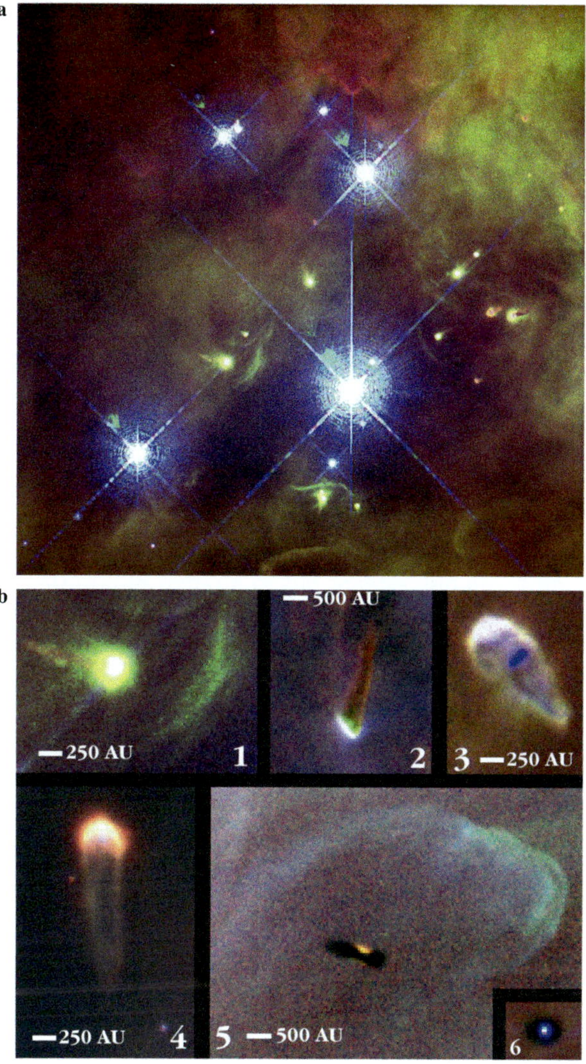

Fig. 6.6 **a** *Hubble Space Telescope* image of the Orion Nebula's blazing heart. Among the *Trapezium* of hot newborn stars, Theta-1 Orionis C to the lower right is largely responsible for transforming previously spheroidal molecular cloud cores into cometary nebulae that stream away from the powerful O-type star. Accompanying the cometary nebulae are curved bow shocks from the star's strong winds. The nebular tails themselves are thought to be the creations of photo-erosive processes that the hot star's intense ultraviolet emission inflicts upon the cloud cores. Inside the cometary nebulae can be found protoplanetary disks, seen in silhouette against the Orion Nebula's fluorescing gases (see Fig. 6.6b) (*credits* NASA,*HST*, C. R. O'Dell and S. K. Wong [Rice University]). **b** Cometary nebula in the center of the Orion Nebula, some of which show protoplanetary disks (proplyds) in edge-on and face-on configurations. The obscuring disks measure several hundred astronomical units (AUs) in diameter, equivalent to that of the remnant debris disk in our own Solar System (*credits* NASA, *HST*, C. R. O'Dell and S. K. Wong [Rice University])

Fig. 6.7 Protoplanetary systems seen face-on and edge-on with respect to our line of sight. *Left*—The protoplanetary disk surrounding the pre-main-sequence star **HL Tauri**. At a distance of only 450 light-years, HL Tau is one of the closest protoplanetary systems to us—thus enabling this remarkably detailed image by *ALMA* at submillimeter wavelengths. The disk extends out to a radius of 80 AU—equivalent to the orbital radii of some Trans-Neptunian Objects in our own Solar System. (Credit: *ALMA* [ESO/NAOJ/NRAO]) *Right*—Infrared image of the Herbig-Haro object **HH 30** showing multiple components from this edge-on perspective, including an obscuring protoplanetary disk, bipolar outflows, and a tail of unknown provenance (*credits JWST*, NASA, ESA, CSA, and R. Tazaki et al.)

During its existence, the disk is a very busy place (see Fig. 6.10). As the planetesimals grow, they collide with one another. There will be winners and losers, as tends to happen in most scenarios of gravitational merging. Moreover, the winners feel the drag imparted by the ambient disk particles. This molasses-like interaction can significantly alter the orbiting radii of the penultimate planets. Indeed, planetary scientists have invoked such a "Grand Tack" to explain the puzzling placement of planets in our own Solar System as well as in those exoplanetary systems that contain giant planets located uncomfortably close to their host stars.

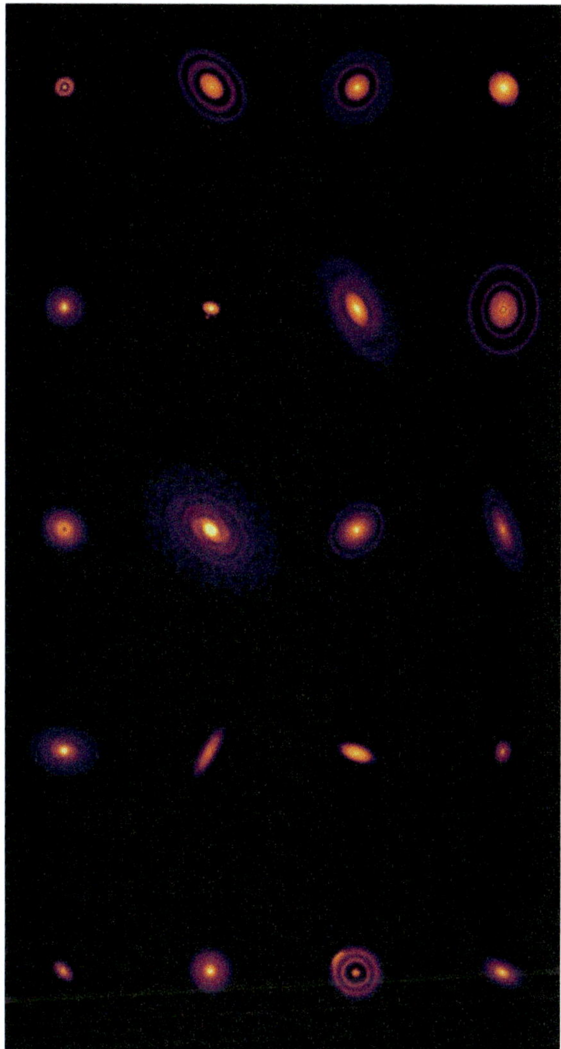

Fig. 6.8 Gallery of 20 protoplanetary disks as imaged at submillimeter wavelengths with *ALMA*. Many of the disks sport concentric rings of emitting dust, while others show spiral structure indicative of density waves coursing through them, and a few of them appear smoothly distributed (*credits ALMA* [ESO/NAOJ/NRAO], S. Andrews et al.; NRAO/AUI/NSF, S. Dagnello)

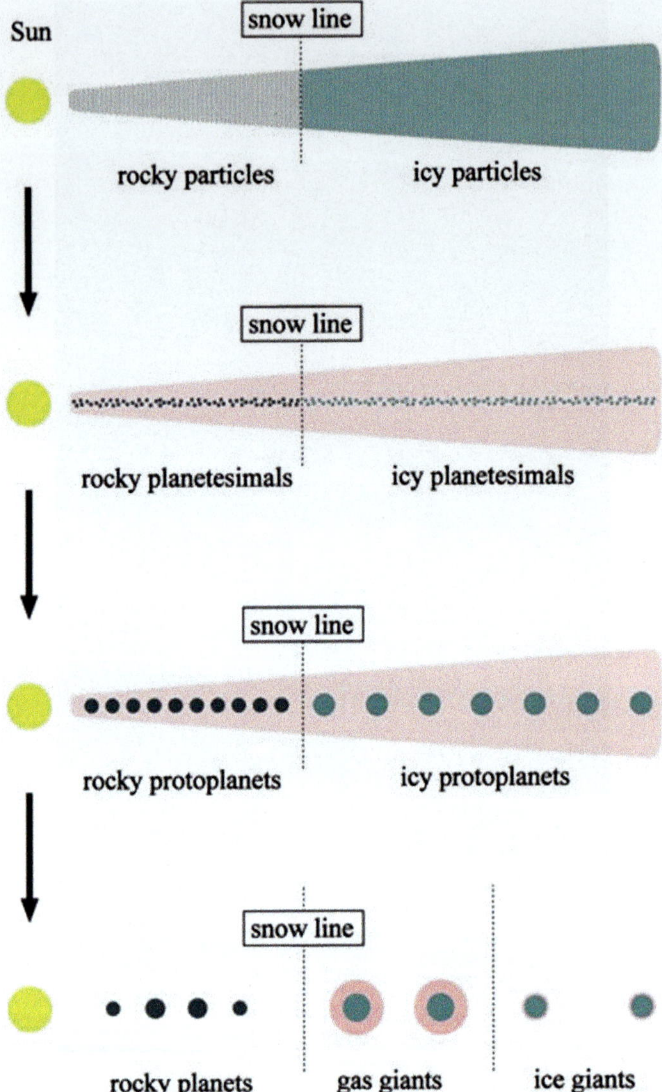

Fig. 6.9 Schematic cross-sections of an evolving protoplanetary system similar to that of the primitive Solar System. The snow lines refer to radii where various gases freeze out, beginning with water vapor. (Credits *Subaru Telescope* / NAOJ)

Fig. 6.10 In this artist's conception, the primitive Solar System featured newborn planets amid a flurry of itinerant debris, some of which pummeled the planets during the first 800 million years of their lives. Once formed, the planets also felt the viscous effects imparted by smaller particles in the disk which served to significantly alter the planetary orbits (*credits* NASA/*FUSE*, Lynette Cook)

7

Biochemical Signatures

What if we had big enough and sensitive enough noses to find out what our Milky Way galaxy smelled like? That might sound like a fool's errand, but we have already made some significant strides courtesy of the meteorites that we have collected here on Earth. Consider the carbonaceous chondrites that comprise about 4.6% of all the meteorites that have impacted Earth. These stony meteorites are rich with tar-like hydrocarbons. If you could get your hands on one of these dark rocks and heat it, you likely would be able to smell the pungent organics.

Our cosmic sensing spree continues with the robotic probes that we have sent to diverse worlds in the Solar System. Both the *Curiosity* and *Perseverance* rovers that are currently roaming the Martian surface are on the hunt for organic compounds. *Curiosity* has already found compelling evidence for benzoic acid ($C_7H_6O_2$) and thiophene (C_4H_4S)—an organic salt that probably got made from solar irradiation of more "normal" organics. *Perseverance* has since detected cyclic molecules involving rings of six carbons each as well as long chain hydrocarbons known as alkanes in an ancient lakebed. These include decane ($C_{10}H_{22}$), undecane ($C_{11}H_{24}$), and dodecane ($C_{12}H_{26}$). Then there is Saturn's largest moon, Titan, with its thick nitrogenic atmosphere. The *Atacama Large Millimeter/Submillimeter Array (ALMA)* of radio telescopes in the Chilean desert has detected the cyclic ring molecule benzene (C_6H_6) along with cyclopropenylidene (C_3H_2)—another

Portions of this chapter first appeared in *The Galactic Inquirer*—a free online journal that covers a diversity of topics in astronomy and space exploration (https://galacticinquirer.net).

W. H. Waller, *Crucibles of Creation*, Astronomers' Universe,
https://doi.org/10.1007/978-3-032-17258-7_7

cyclic molecule that is highly reactive. The *Cassini* orbiter further confirmed a variety of nitrogenous molecules mingling with the methane (CH_4) and ethane (C_2H_6) vapors that are thought to condense, rain down, and pool on the surface of Titan.

All these planetary explorations beget further questions regarding the chemical pathways that could have led to life on Earth—and on other worlds as well—where Astrochemistry turns into Astrobiology (cf. Ehrenfreund and Charnley 2000; Kolb 2019; Donagal-Goldman and Wright 2021). But much can be learned by exploring the much larger *galactic ecosystems*—the subject of the present book—as they have been found to be rich repositories of organic chemicals and complex chemistries. Herein, I review the astrochemistry associated with galactic ecosystems—first focusing on cold-dark molecular clouds while setting the stage for assaying highly-energized HII regions and starbursts in Chap. 10. These sundry regions play host to diverse dynamical and chemical processes which beg understanding.

But first, you might be wondering how any sort of chemistry can take place in the near-vacuum of interstellar space. Chemistry involves atoms finding one another and bonding to form molecules. Therefore, the rarefied gases contained within galactic ecosystems would appear to be the least amenable to such socializing. All is not lost, however, as the dust grains that permeate the galactic ecosystems provide handy platforms for itinerant atoms to settle upon and so make the necessary hook-ups. The various ices that coat these dust grains provide further fodder for fascinating chemistries to ensue. Since the 1970s, spectroscopic observations have emphatically shown that galactic ecosystems harbor rich brews of complex molecules.

A Molecular Census

Most molecules announce their presence by changing their rotational and/or vibrational energy states and so radiating in the radio, microwave, and infrared parts of the electromagnetic spectrum. Favorite targets of discovery include the Sagittarius B2 molecular cloud near the Galactic center, the circumstellar environment of CW Leonis, and the nearby Taurus Molecular Cloud. Indeed, the TMC has provided one in three of all molecular detections in space. To date, more than 300 unique molecules have been discovered within these sorts of interstellar and circumstellar environments—each molecule containing from two to more than ten atoms (see Table 7.1 and https://en.wikipedia.org/wiki/List_of_interstellar_and_circumstellar_molecules#Molecules).

Table 7.1 Abridged listing of interstellar molecules as observed at centimeter, millimeter, and sub-millimeter wavelengths

		2 Atoms		
Hydrogen (H_2) Nitric Oxide (NO) Carbon Monoxide (CO)	Deuterized Hydrogen (HD) Nitrogen Sulfide (NS) Carbon (C_2)	Hydroxyl Radical (OH) Silicon Sulfide (SiS) Carbon Monosulfide (CS)	Silicon Monoxide (SiO) Methylidyne Radical (CH)	Sulfur Monoxide (SO) Cyanogen Radical (CN)
		3 Atoms		
Water (H_2O) Ethynyl Radical (C_2H) Dicarbon Sulfide (C_2S)	Heavy Water (HDO) Hydrogen Cyanide(HCN)	Hydrogen Sulfide (H_2S) Deuterium Cyanide (DCN)	Sulfur Dioxide (SO_2) Formyl Radical (HCO)	Nitroxyl (HNO) Dicarbon Monoxide (C_2O)
		4 Atoms		
Ammonia (NH_3) Isothiocyanic Acid (HNCS)	Formaldehyde (H_2CO) Propynylidyne (C_3H)	Isocyanic Acid (HNCO) Cyanoethynyl (C_3N)	Thioformaldehyde (H_2CS) Tricarbon Monoxide (C_3O)	Acetylene (HC_2H) Tricarbon Sulfide (C_3S)
		5 Atoms		
Methane (CH_4) Ketene (CH_2CO)	Cyanamide (H_2HCN) Butadinyl (C_4H)	Formic Acid (HCOOH) Methylenimine (CH_2NH)	Cyanoacetylene (HC_3N) Cyclopropenylidene (C_3H_2)	Cyanomethyl (CH_2CN)
		6 Atoms		
Methyl Alcohol (CH_3OH)	Formamide ($HCONH_2$)	Methyl Mercaptan (CS_3SH)	Methyl Cyanide (CH_3CN)	Pentynylidyne (C_3H)
		7 Atoms		
Acetaldehyde (CH_3CHO) Vinyl Cyanide (CH_2CHCN)	Methylamine (NH_2CH_3)	Methylacetylene (CH_3C_2H)	Cyanodiacetylene (HC_3N)	Hexatrinyl (C_6H)
		8 Atoms		
Methyl Formate ($HCOOCH_3$)	Methyl Cyanoacety-lene (CH_3C_3N)	Acetic Acid (CH_3COOH)	Glycoaldehyde ($H_2COHCHO$)	
		9 Atoms		
Ethyl Alcohol (CH_3CH_2OH)	Ethyl Cyanide (CH_3CH_2CN)	Dimethyl Ether (CH_3CH_3O)	Methyl Diacetylene (CH_3C_4H)	Cyanohexatriyne (HC_7N)
10 Atoms	**11 Atoms**	**12 Atoms**	**13 Atoms**	**24 Atoms**
Glycine (NH_2CH_2COOH)	Cyanooctatetrayne (HC_9N)	Benzene (C_6H_6)	Cyanopentaacetylene ($HC_{11}N$)	Anthracene (PAH) ($C_{14}H_{10}$)

As the above listing shows, many of the interstellar and circumstellar molecules are "organic," having carbon atoms as parts of their structures. So, we can already see that the chemical species found in galactic ecosystems provide some of the essential feedstock for developing the more complex molecules characteristic of life on Earth such as lipids, carbohydrates, amino acids, and nucleotides. Further chemical processing can ensue within the denser protoplanetary disks and, ultimately, upon the surfaces of planets that coalesce from the disks.

The smaller molecules in the listing include water (H_2O), formaldehyde (H_2CO), ammonia (NH_3), methane (CH_4), and hydrogen cyanide (HCN)—all of which are vital to various biotic processes on Earth. HCN, in particular, is thought to be a key ingredient for synthesizing amino acids and other macromolecules that make up what we regard as biological systems (see Fig. 7.1).

Of the latter big molecules in the listing, the most familiar include ethyl alcohol (CH_3CH_2OH) with its pungent aroma and psychoactive quaff that many people nonetheless enjoy, acetone ($(CH_3)_2CO$) with its fruity but toxic fumes, ethyl formate (C_2H_5OCHO) that smells like rum, the sugar glycoaldehyde ($C_2H_4O_2$), the cyclic ring molecule benzene (C_6H_6), and all the polycyclic aromatic hydrocarbons (PAHs) that are made from weaving benzene into pliable molecular mats. PAHs provide the characteristic aromas that you can sense from a charred burger or the backend of a diesel bus (see Fig. 7.2). We can now see that molecular clouds like the TMC, along with their pre-planetary progeny, appear optimized for synthesizing all sorts of complex molecules—including those associated with pre-biotic chemistry.

Glaring in their rarity in the above listing are molecules that include phosphorus—a key component in macromolecules such as DNA, RNA, phospholipids, as well as the adenosine triphosphate (ATP) molecules which serve as energetic "batteries" within cells. More comprehensive listings include detections of phosphorus nitrate (PN) and phosphorus monoxide (PO) within molecular clouds along with methinophosphide (HCP), cytochrome-c peroxidase (CCP), and phosphine (PH_3) that have been found in circumstellar environments (Fontani 2024). But these detections were hard won. Phosphorus represents one of the vital elements for life on Earth, sharing the CHNOPS acronym with the other 4 elements essential to terrestrial life. Yet it is puzzlingly rare in space, having 1/100th the abundance of sulfur, and 1/1000 that of carbon. Until we better understand this biogenic "bottleneck," we will have to entertain the possibility of other chemical pathways generating what we would regard as biological capabilities.

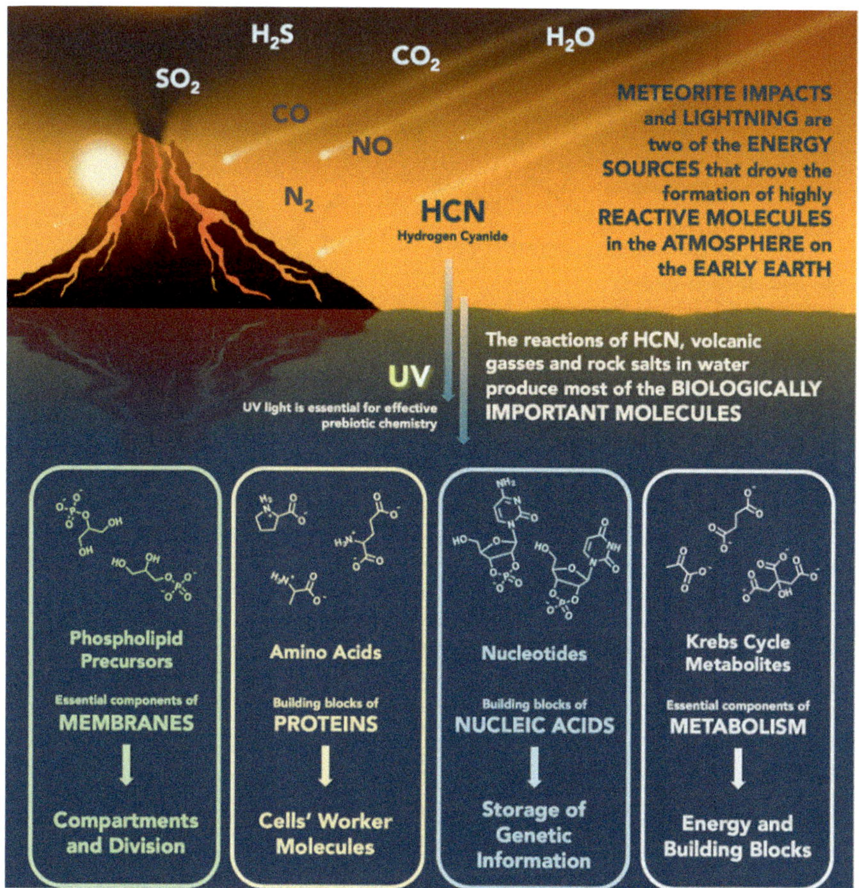

Fig. 7.1 The unique bonding configuration (and electrical polarity) of hydrogen cyanide (HCN) makes it amenable to building much larger macromolecules, including the lipids, amino acids, nucleotides, and various metabolites that we have on Earth today. In this artistic depiction, the water on primitive Earth serves as the key chemical solvent for developing these biological precursors (*credits* Courtesy of Angelica Mariani with reference to B. H. Patel et al. 2015, https://pmc.ncbi.nlm.nih.gov/articles/PMC4568310/)

Growing Chemical Complexity

Recent theoretical and laboratory work has shown how the physical conditions in galactic ecosystems can drive chemical activity and so build-up chemical complexity. Ultraviolet radiation from embedded hot stars, in particular, can energize simple molecules found in interstellar ices, thus producing a rich variety of chemical species in a process known as *photolysis* (Oberg 2016, 2023). The resulting organic species include carbon monoxide (CO),

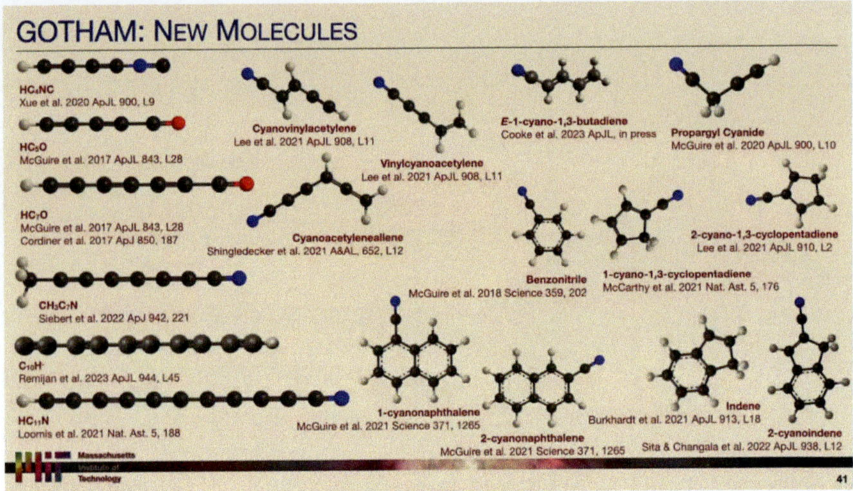

Fig. 7.2 Depictions of long-chain (aliphatic) and cyclic ring (aromatic) molecules that have been recently detected at radio wavelengths in the Taurus Molecular Cloud with the *Green Bank Telescope*. Some of the long-chain molecules are fragile and so can only be found in small and cold molecular clouds such as the TMC, where the disruptive effects of particle collisions and UV radiation are minimal. By contrast, the cyclic molecules are far more durable thus enabling their persistence in much harsher environments that are energized by massive stars (*credit* Courtesy of Brett McGuire [MIT/NRAO])

formaldehyde (H_2CO), methane (CH_4), methanol (CH_3OH), dimethyl ether (CH_3OCH_3), methyl formate ($HCOOCH_3$), and ethylene glycol $(CH_2OH)_2$ as obtained via diverse pathways (see Fig. 7.3). But even in clouds that lack hot stars, similar photolytic processing can occur on the scale of the protoplanetary disk, where the host star is near enough to energize the ices and gases contained within the disk.

Recent laboratory experiments have simulated such physical conditions, whereby UV radiation energizes various ices that are situated in an evacuated chamber. The resulting chemical products obtained by these experiments have provided encouraging proofs of concept for photolytic processing as an important agent in fostering ever more complex chemistries.

We will revisit this rapidly evolving field of astrochemistry in Chap. 10, where we will consider much larger galactic ecosystems that have been energized by caches of hot UV-bright stars along with intense winds from these stellar behemoths and shock waves from their explosive deaths. What might transpire in these starbursting realms remains fairly speculative but well worth our consideration.

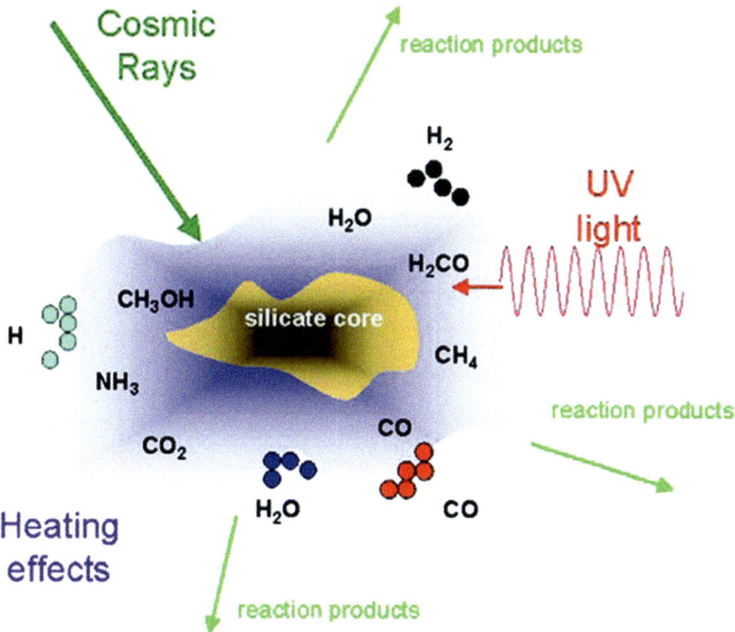

Fig. 7.3 A model of chemical reactants and products occurring within the icy mantle of an interstellar grain. Besides heating of the grains and the energizing influence of cosmic rays, photolysis by UV irradiation is thought to play a major role in driving ever greater chemical complexity (*credit* Royal Society of Chemistry)

Part III

Spawning in the Spotlight—Massive Stars and Their Nebular Impacts

I know not how to describe it [the great nebula in the constellation of Orion] better than by comparing it with a curdling liquid, or a surface strewed over with flocks of wool, or to the breaking up of a mackerel sky, when the clouds of which it consists begin to assume a cirrus appearance.

—John Herschel, in *Results of Astronomical Observations Made During the Years 1834–1838 at the Cape of Good Hope* (1847 CE)

8

Content, Structure, and Dynamics of HII Regions

Imagine visiting a nursery ward in an urban hospital, where before you are rows upon rows of newborns swaddled in cozy blankets within their personal cubbies. You learn from their charts that these babies typically weigh about 7 pounds. A small fraction of them weighs half that amount, while another small cohort weighs up to twice the average. So, you can see that the total range of newborn weights varies by a factor of no more than four.

Now, consider the range of masses for stars at their birth. The lower mass limit is thought to be 1/10 that of the Sun. Any lower the mass, the object cannot muster the core temperatures necessary to fuse hydrogen into helium, and so is regarded as a brown dwarf rather than a bona fide star. At the high-mass end, stars 100 times more massive than the Sun have been surmised from their spectra and other observables. The theoretical upper-limit of about 150 solar masses is set by radiation pressure from these luminous powerhouses (see Appendix E.7). So, we find ourselves dealing with a factor of approximately 1000 in mass between the puniest and most ponderous stars. The temperatures on the stellar "surfaces" also range drastically—from around 3000 K for the lowest-mass M-type stars to about 60,000 K for the most massive O-type stars. That means the lowest-mass stars radiate mostly in the infrared part of the electromagnetic spectrum, while the highest-mass stars radiate mostly in the ultraviolet. This wide range of mass-dependent surface temperatures and corresponding spectral energy distributions has a strong effect on the resulting luminosities. Along the main sequence, astronomers have found that the stellar luminosities vary roughly as the 3rd or 4th power of the mass. That means the total range in luminosities is more like a factor of a billion. Yikes! (see Fig. 8.1).

W. H. Waller, *Crucibles of Creation*, Astronomers' Universe,
https://doi.org/10.1007/978-3-032-17258-7_8

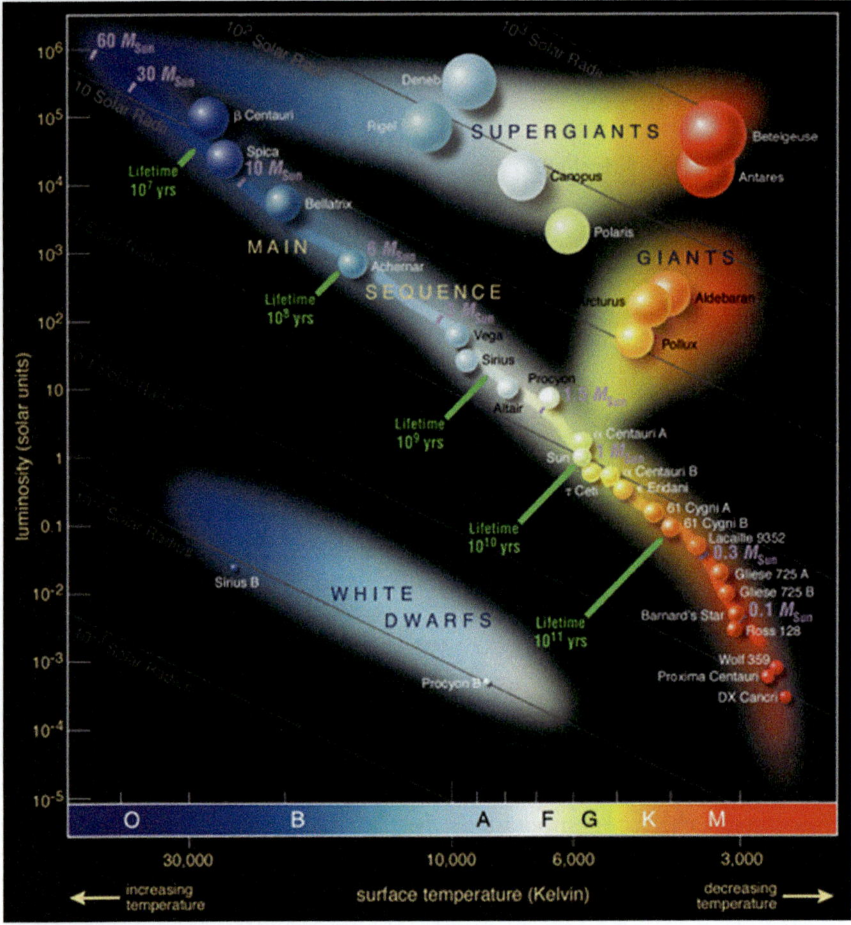

Fig. 8.1 In this schematic Hertzsprung-Russell diagram, the Main Sequence traces those stars that are actively fusing hydrogen into helium in their thermonuclear cores. Stars typically spend 90% of their lives in this evolutionary phase. Along the Main Sequence, the stellar surface temperatures and luminosities vary drastically. The most massive stars display searing surface temperatures, thus producing prodigious amounts of ultraviolet emission. They blaze forth with luminosities that exceed those of the least massive stars by a factor of a billion (*credit* European Southern Observatory)

A more careful examination of the stellar properties also reveals startling variations in the stars' overall lifetimes (see also Appendix F). While the Sun has an expected hydrogen-fusing lifetime of 10 billion years, the most massive stars are thought to last only a few million years. "Live fast and die young" pretty much characterizes these superstars. Besides having severely truncated lifespans, the most massive stars are also exceedingly rare. For every 100 stars of solar-mass that get forged within a galactic ecosystem, you might find only one star with a mass greater than 10 Suns. That is why most massive stars are found within young populous clusters that are still embedded within their natal birth clouds. Herein, we can bear witness to drastic transformations afoot.

Ionizing Radiation from Massive Stars and Its Nebular Consequences

In their fleeting moments on the cosmic stage, massive stars inflict irreversible changes upon their nebular environments (see Appendix E.4). Let's begin with their intense emission at ultraviolet wavelengths. Once the thermonuclear fires ignite in their cores, massive stars attain surface temperatures exceeding 25,000 K. That's hot enough for most of the stellar emission to occur at ultraviolet (UV) and extreme ultraviolet (EUV) wavelengths. Our human eyes can't see most of what these stars are actually putting out! For hydrogen molecules in any surrounding nebulae, dissociation will occur at UV wavelengths less than 253.7 nm. When the H_2 molecules absorb such high-energy photons, their electric bonding will be insufficient to hold them together, and so they will break-up into their constituent H atoms. These H-atoms, in turn, will be subject to getting ionized by any stellar EUV photons with wavelengths less than 91.2 nm. Each H-atom's single electron will get ripped away, thus producing a plasma of free protons and electrons. We now have the basics of an HII region.

Astronomers can map these HII regions courtesy of their fluorescent emission at particular wavelengths. The most commonly detected emission is from ionized hydrogen as it re-captures a previously free electron to again form a neutral H-atom. The recombined electron cascades down a staircase of quantum energy levels to ultimately reach its ground state. One of these jumps is from the 3rd energy level above the ground state to the 2nd level. The resulting emission of excess energy occurs at the distinctly red wavelength of 656.3 nm. You have already seen this so-called H-alpha emission in the visible-light views of the Milky Way (see Figs. 3.5–3.8). But HII regions also glow at other discrete wavelengths. First, the ionized hydrogen produces a whole series of fainter emission lines with teal, blue, and violet colors resulting from the various jumps of the recombined electron as it cascades down the energy levels. Second, hydrogen is not alone within the HII regions. A rich medley of emission lines arises from the ionization and excitation of other elements. These often include neutral oxygen [OI], singly ionized oxygen

[OII], singly-ionized nitrogen [NII], singly-ionized sulfur [SII], and doubly-ionized oxygen [OIII]. Here, the square brackets denote electron transitions that are rare (or "forbidden") at densities typical of Earth's atmosphere but nonetheless occur at the relatively low densities extant in HII regions (see Fig. 8.2).

Sometimes full-color images of HII regions can show how the photoionization is proceeding. The blue-green [OIII] emission often occurs closer to the hot ionizing stars, where the most energetic EUV photons get absorbed first. Farther out, the H-alpha emission typically dominates, as the once atomic hydrogen gets ionized by the remaining EUV photons. And even beyond the H-alpha emitting surface of the photoionized nebula, one can catch the [SII] emission ramping up along the ionization shock front. These myriad effects can be effectively visualized by imaging the nebula through narrowband filters that pass the specific line emission of interest and then combining the separate monochromatic images into one color-coded composite image (see Fig. 8.3).

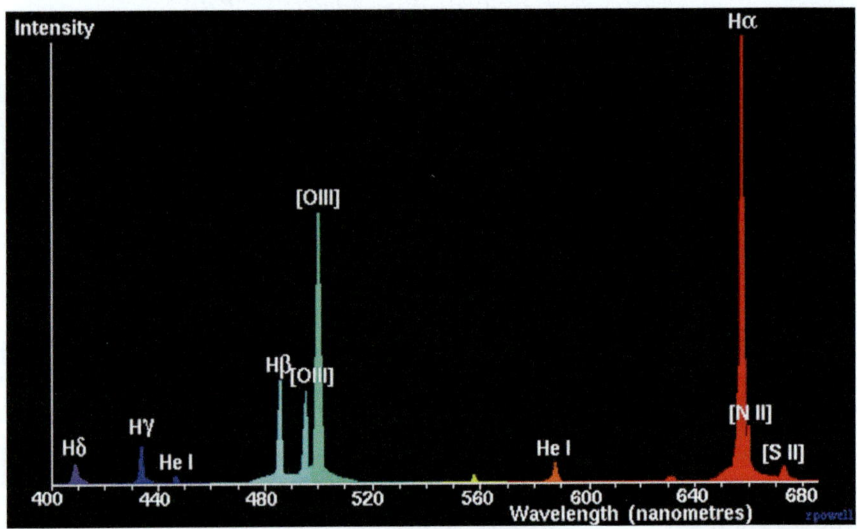

Fig. 8.2 The visible spectrum of a filament associated with the Carina Nebula (NGC 3372) shows multiple emission lines of hydrogen, sulfur, nitrogen, helium, and oxygen in various ionization states (*credits* From R. Powell's *Atlas of the Universe* at: http://www.atlasoftheuniverse.com/nebulae/ngc3372.html with reference to C. Hua and A. Llebaria (1981), "Optical Spectrum of the Filamentary HII Region North of the Carina Complex," *Astronomy and Astrophysics*, Vol. 94, p. 12)

Fig. 8.3 The **Rosette Nebula** (NGC 2244/2237) in the constellation of Monoceros was imaged through separate narrowband filters that pass [OIII], H-alpha, and [SII] emission. These separate images were then color-coded blue, red, and yellow, respectively. Here, the blue-coded [OIII] emission is most prominent towards the center of the nebula, where the exciting hot stars are clustered. The hydrogen and sulfur atoms take less energy to get singly ionized compared to the twice-ionized oxygen. Therefore, the red-coded H-alpha and yellow-coded [SII] emission trace the lower-ionization states of the respective species and so help to delineate the outer working surfaces of the photo-ionized nebula. This HII region is roughly 100 ly in size (*credits* Courtesy of Adam Block and Russ Croman. https://www.adamblockphotos.com/uploads/4/4/2/5/44258693/rosette_hso_final_croman.jpg)

Let the Scorching Begin

The photo-ionizing process depends on the number of hot massive stars in any cluster, the density of gas in the surrounding birth cloud, as well as the quantity of dust that competes with the gas for the ionizing photons. An equilibrium of sorts readily plays out between the rate of hydrogen photo-ionizations and the rate of recombinations back to the atomic state. This balance ultimately determines the size of the resulting HII region (see Appendix E.5).

For a single O-type star irradiating a cloud with an average density of 100 H-atoms per cubic centimeter, simple modeling would predict an HII region with a diameter of about 20 ly. Addition of dust would reduce this size by no more than half. These estimates come close to approximating the

Orion Nebula, whose actual size is about 13 ly (see Fig. 8.4). Much larger star clusters with 10–100 O-type stars can carve out HII regions with sizes of 50–100 ly—what we would regard as true giants of the genre.

Fig. 8.4 The **Orion Nebula** (M42, NGC 1976) is the closest example of burgeoning star formation and consequent nebular disruption in our part of the Galaxy. This HII region is located on the near surface of the Orion Giant Molecular Cloud, itself containing close to a million suns worth of cold molecular gas. The fluorescing nebula subtends an angle of about a half-degree in the sky and so can be readily viewed with binoculars and small telescopes. At a distance of 1300 ly, it has a true size of about 13 ly, thus qualifying it as a "classical" HII region. At the core of M42, the Trapezium Cluster of hot stars blazes forth (see also Fig. 6.6). One O6-type star in the cluster, Theta-1 Orionis C, appears to dominate the photoionization of the nebula. Along the periphery of the nebula, one can trace the remnant bubble of expanding ionized gas. This image is a composite of broadband and narrowband exposures made by the *Hubble Space Telescope* and the European Southern Observatory's *2.2-m telescope* (*credits* NASA, ESA, Massimo Robberto [STScI, ESA], Hubble Space Telescope Orion Treasury Project Team)

Sometimes the photo-ionizing process meets resistance in the form of denser regions, whose entrained dust blocks further photo-excavations. Like the "hoodoos" in Bryce Canyon and other eroded landscapes on Earth, these so-called "pillars of creation" manifest the dramatic transfigurations of nebular matter that are underway in some HII regions (see Fig. 8.5).

Thermal Energetics

Once cold molecular gas gets photo-dissociated and photo-ionized, the resulting plasma of charged particles heats up to temperatures of 7000–10,000 K. And like any heated gas, it exerts excess pressure on its surroundings. Soon, the ionized gas expands and so augments the cavitation already underway in the hosting cloud.

Sometimes, the expanding gas breaks through the cloud and so produces a hot ionized wind. This could be the case with the Orion Nebula, where we appear to be looking through a breach in the bubble that was inflated by the Trapezium Cluster (see Fig. 8.4). Once the ionized gas escapes, it is free to diffuse into the ambient interstellar medium. We now have a mechanism for mixing things up on the galactic stage. In the next chapter, we will consider other mechanisms for injecting matter and energy into (and beyond) galactic ecosystems.

a

Fig. 8.5 **a** The Eagle **Nebula** (M16, NGC 6611) in the constellation of Serpens is a churning region of clustered star formation, whose hot massive stars have carved out an irregular cavity filled with ionized gas. This color composite image was made from three separate exposures through blue, green, and red broadband filters and then combined with the intent of emulating a "natural" palette. In the image one can see several "pillars" of dense gas and dust resisting the radiative onslaught from the hot stars. The total scene spans about 60 ly at an estimated distance of 5700 ly (*credit* European Southern Observatory via Wikipedia and Creative Commons Attribution 4.0 International License. http://creativecommons.org/licenses/by/4.0/). **b** Closeups of a pillar in M16 as imaged through narrowband filters with the *Hubble Space Telescope* and color-coded with [OIII] emission in blue, H-alpha emission in green, and [SII] emission in red. These images show the nebular effects of intense UV irradiation by the hot ionizing stars. Streams of excited gas can be seen flowing away from the pillars, exposing "evaporating gaseous globules (eggs)" in the process. Such photoerosive dynamics could prove fatal to any condensing cloud cores in the eggs (*credits*: NASA, ESA, *HST*, and the Hubble Heritage Team [STScI/AURA])

Fig. 8.5 (continued)

9

Nebular Energetics and Their Powering

In the previous chapter, we focused on the intense ultraviolet radiation that hot massive stars emit and the dramatic effects that this radiation inflicts upon the stars' nebular birthplaces. But UV light is not the only energetic emanation expressed by these stars. Over their brief lifetimes, massive stars blow strong winds that can buffet and shock their birth clouds. Wolf-Rayet stars and other evolved exemplars of the massive stellar genre blow especially powerful winds, with mass-loss rates exceeding 10^{-5} suns per year worth of outflowing material. After a few million years, the cores of the most massive stars will run out of fusible fuel and collapse. The outer layers of these stars will then respond by exploding as Type II supernovae. The resulting blast waves can collisionally energize the ambient gas to temperatures of several million degrees, thus ramping up the gas pressures and so inducing further evacuation of the remnant nebula—and perhaps even generating a starburst wind. A Sunday comics rendering of such a starbursting scene is shown in Fig. 9.1, where a giant molecular cloud has generated a populous cluster of stars whose most massive stars are both irradiating and scouring their surroundings. Therein, a recent supernova explosion has begun to evacuate a region filled with tenuous hot gas that will continue to expand into the cloud and perhaps escape as a starburst outflow.

W. H. Waller, *Crucibles of Creation*, Astronomers' Universe,
https://doi.org/10.1007/978-3-032-17258-7_9

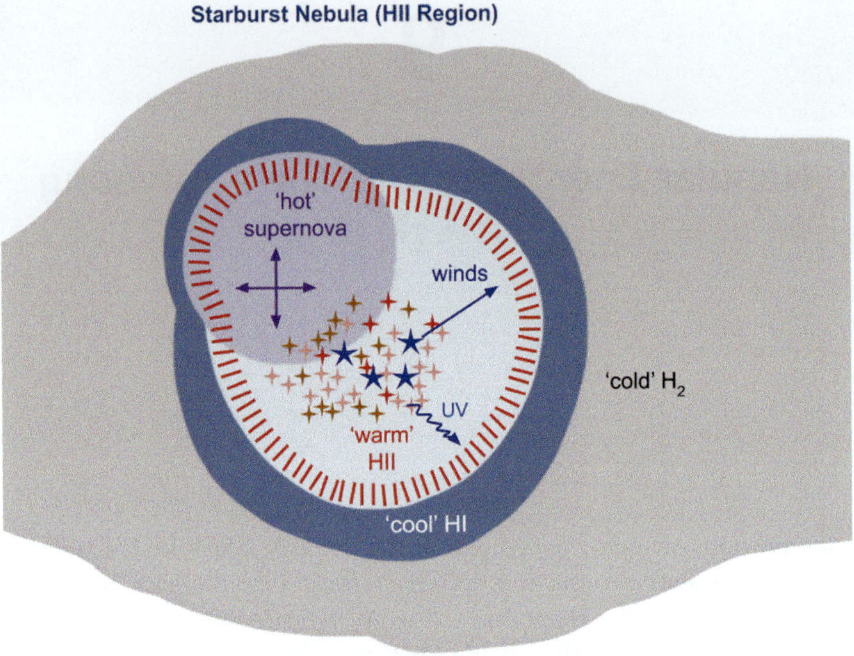

Starburst Nebula (HII Region)

Fig. 9.1 Cartoon "snapshot" of a populous star cluster and its energetic consequences within a galactic ecosystem. Among the many newborn stars in the cluster, a few of the stars are massive and hot enough to photo-dissociate and then photo-ionize the surrounding birth cloud of molecular hydrogen, thus creating an HII region of "warm" 10,000 K gas. The thin skin, where molecular hydrogen transitions into atomic and then ionized phases is known as a photo-dissociation region (PDR). Powerful winds from the massive stars further evacuate the ionized cavity and energize its working surfaces. One of the massive stars has recently exploded as a supernova, shock-heating the gas to millions of degrees Kelvin. The excess pressure of this superheated gas causes further expansion and perhaps even a starburst wind (*credits* W. H. Waller and L. H. Slingluff)

Thar She Blows!—Stellar Winds and Their Energizing Effects

Even before massive stars begin fusing hydrogen in their cores, they generate strong outflows akin to those first discussed in Chap. 6 (where we focused on young stellar objects of lower-mass). There, the outflows were highly collimated in the form of jets directed perpendicular to the pre-planetary disks. Evidence for similar behavior among the massive stellar elite can be found in the Rho Ophiuchi Nebula, Orion Nebula, Carina Nebula, and other massive star-forming regions. In general, outflow power increases with the stellar source's mass and luminosity. So, it is worth our while to investigate the winds generated by the most massive stars along with their environmental impacts.

Much of what we know about stellar winds from massive stars comes from those few massive stars that are sufficiently clear of their nebular birthsites to be observable at visual and ultraviolet wavelengths. The resulting spectra often show absorption lines that are blue-shifted relative to the corresponding emission lines. The line emission is thought to arise from the circumstellar nebula associated with the winds. The absorption lines represent that portion of the nebula that lies between the star and us. According to the Doppler effect, their blue-shifts indicate flows in our direction. By measuring the degree of blue-shifting of the absorption lines relative to the emission lines, astronomers can estimate the associated outflow velocities (see Fig. 9.2).

To ascertain the mass-loss rates from these windy stars, it is necessary to also observe the radio continuum emission produced by the outflowing ionized winds. Knowledge of the star's distance plus the observed radio flux

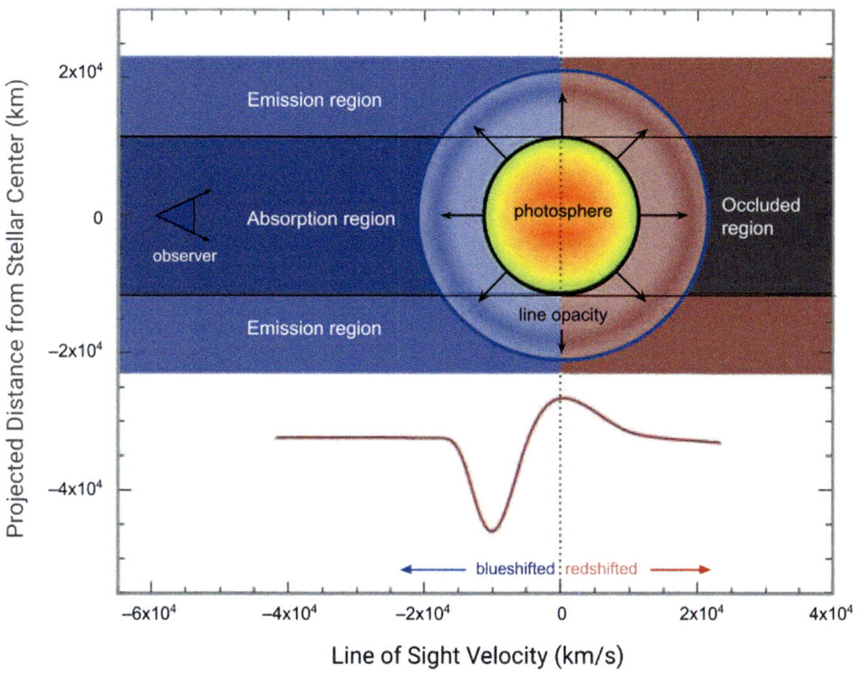

Fig. 9.2 Generic P-Cygni spectral profile and corresponding geometry of a hot star surrounded by an expanding wind nebula. The P-Cygni moniker refers to the prototype supergiant star that was first observed to display such spectral features (*credit* Adapted with changes from figure by Daniel Kasen, U. California Berkeley, https://supernova.lbl.gov/~dnkasen/tutorial/p-cygni/pcygni_form.jpg)

will yield the actual luminosity of the radio continuum emission. This radiative luminosity along with an appropriate conversion of the radio flux into a nebular mass and mass density, can then be combined with the outflow velocity to obtain the mass-loss rate via the following formula (Lozinskaya 1992).

$$dM/dt = 4\,\pi\,r^2\,\rho(r)v(r),$$

where dM/dt is the mass loss rate, $\rho(r)$ is the mass density at radius r, $4\,\pi\,r^2\,\rho(r)$ is the mass per unit length at radius r, and v(r) is the outflow velocity (see Appendix E.6 for a more detailed derivation). The resulting mass-loss rates are shown in Fig. 9.3.

For the massive stars, the intense winds arise from the stars' profligate flux of UV photons interacting with hydrogen, carbon, silicon, nitrogen,

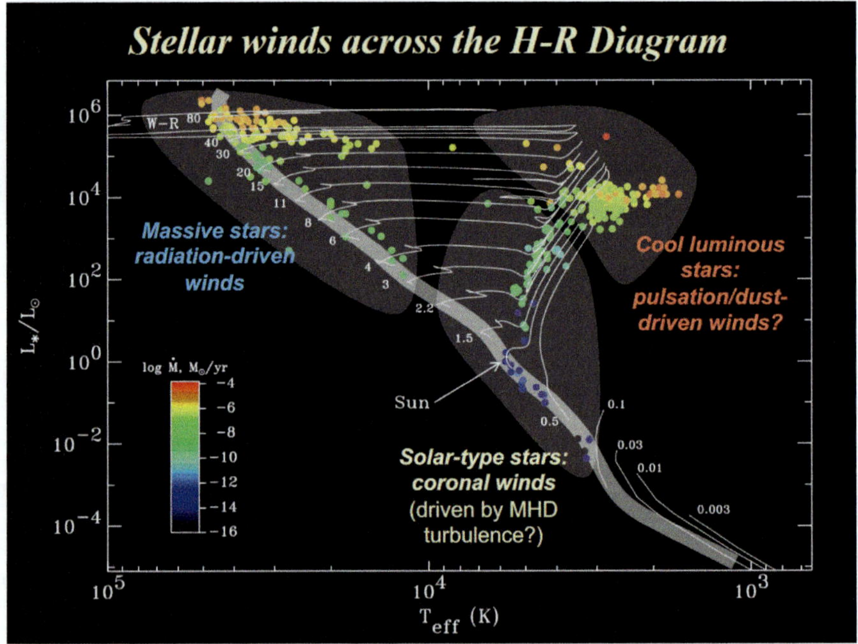

Fig. 9.3 Hertzsprung-Russell (H-R) diagram of stellar luminosity vs. surface temperature with stars of different mass populating it. The varying colors indicate differing mass-loss rates that result from the stars' respective winds. Sun-like stars have very low mass-loss rates of ~ 10^{-14} solar masses per year, while massive stars and lower-mass red-giant stars have mass-loss rates as high as ~ 10^{-5} solar masses per year. During their lifetimes, these stellar blowhards can exhaust up to 50% of their original masses (*credit* Courtesy of S. R. Cranmer, Laboratory for Astronomy and Space Physics, Colorado University)

oxygen, and other ionized atoms in the stellar atmospheres. As the atoms absorb photons of particular wavelengths, they receive a boost in momentum conveyed by the photons. This so-called radiation pressure scales with the star's luminosity, which for the hot massive stars, can drive a wind at velocities of several thousand km/s (see Appendix E.7). The corresponding mechanical luminosity of these winds is readily derived using the common formula for kinetic energy (KE = ½ M v^2), producing

$$L_W = {}^1/_2(dM/dt)v^2.$$

For the highest-mass stars, these mechanical luminosities can rival their radiative counterparts (see Table 9.1). The massive stars that have evolved from their main-sequence phase (luminosity class V) to become supergiants (luminosity class I), O_f and Wolf-Rayet (WR) stars are especially powerful, blowing the equivalent of 100,000 suns worth of windy luminosity.

Table 9.1 Stellar wind energetics

Type of star (or association)	Mass-loss rate (solar masses/ yr)	Wind velocity (km/s)	Wind power (L_W/ L_{Sun})	Duration of wind (yr)	Total wind energy (E/E_{SN})	Size of impacted region (ly)
O9 V	10^{-8}	1000	~ 1.0	10^7	~ 10^{-3}	30–150
O5 I	10^{-4}	3000	~ 10^5	10^5	~ 1	30–150
O_f/WR	5 × 10^{-5}	2000–3000	10^4 to 10^5	10^5	10^{-2} to 10^{-1}	30–150
Central star, planetary nebula	10^{-7}	3000	~ 100	10^4	10^{-4}	0.3–3.0
Red giant	10^{-6}	10	~ 0.1	10^6	10^{-5}	30
OB association	10^{-4}	2000–3000	~ 10^5	10^7	~ 60	300–3000
Supernova	0.1–1.0	~ 10,000	~ 10^5	10^4 to 10^5	1.0	30–300

Notes to table: The O_f stellar type refers to hot and windy stars that still retain their hydrogen atmospheres. The WR type refers to stars that have lost their hydrogen atmospheres thus exposing their even hotter interiors of helium. Planetary nebulae are produced by Sun-like stars at the end of their lives. The wind power, L_W, is referenced to the *radiative luminosity* of the Sun (4.0 × 10^{33} erg/s). A supernova explosion lasts no longer than a year, but its evolving remnant can last 10^4 to 10^5 yr, as listed in the table. The estimated total energy of a supernova is about 10^{51} ergs, or 10^{44} Joules

Adapted from Table 19 in T. A. Losinskaya 1992

Here, the types of massive stars range from hot OV-type stars that are still (barely) on the Main Sequence, evolved OI-type supergiant stars that have expanded to prodigious sizes, and even stranger O_f and Wolf-Rayet (WR) stars whose mighty winds have exposed their exceptionally torrid interiors. Surface temperatures as high as 200,000 K characterize many of the WR stars. Often surrounding these massive misfits, one can find nebulae that they have blown with their powerful winds (see Fig. 9.4). The nebulae contain warm grains of dust that glow at near- and mid-infrared wavelengths. They also sport gaseous components that fluoresce from the combined UV irradiation and windy bombardment—showcasing a medley of emission lines from multiply-ionized atoms. The evolutionary path from the Main Sequence remains controversial, but it is thought that the O_f and WR phases represent an evolutionary sequence from the most massive main sequence stars, on their way to exploding as Type II supernovae.

End of the Line—Supernovae and Their Energetic Remnants

Throughout their lives, stars must vigorously oppose the inward directed gravitational forces that would otherwise cause them to collapse under their own weight. By fusing light elements into increasingly heavier elements, they can exert thermal pressures and so stave off the inevitable collapse. However, once a massive star has fused in its core a succession of elements progressing from hydrogen to helium, carbon, neon, oxygen, silicon, and then to iron, the game is up. Iron requires more energy to fuse than it can dish out. The thermonuclear core immediately goes dormant and collapses in less than a second. The subsequent release of gravitational energy first manifests as a flood of neutrinos which then drives the hapless star to explode big time.[1]

What remains of the once mighty star is an incredibly compact neutron star (the size of a city) or an unfathomable black hole, depending on the mass of the collapsing core. Meanwhile, the star's outer layers have blasted away to form a highly-ionized nebula with temperatures in the millions of Kelvins.

[1] Core-collapse supernovae of Type II are but one type of supernova. They typically show line emission from hydrogen resulting from the stellar layers of hydrogen that were expelled in the explosion. Type Ia supernovae do not show hydrogen in their spectra, as they result from the obliteration of a white dwarf that has already lost its outer layers of hydrogen. The white dwarf is prompted to explode, when it has accreted enough mass from a close companion star to exceed the Chandrasekhar limit of 1.4 Suns. Because Type Ia supernovae all have progenitor masses near the Chandrasekhar limit, they have fairly constant luminosities. This attribute enables them to be used as "standard candles" for fathoming the distances to galaxies that host them – key measurements for determining the expansion rate of the Universe (the so-called Hubble constant).

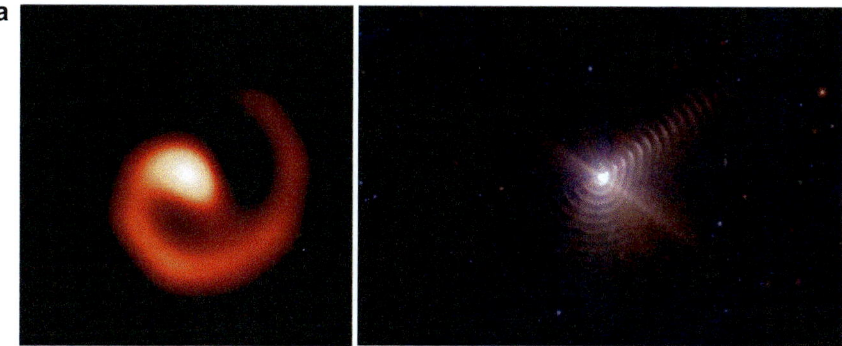

Fig. 9.4 **a** Dusty ejecta from Wolf Rayet stars. *Left*—Located in the constellation of Sagittarius at a distance of 8400 ly, **WR 104** consists of three stars, two of which are closely orbiting their mutual center of mass. The combined winds from the true WR star and its OB-type companion have produced a dusty outflow that spirals outward like a stream of water from a spinning lawn sprinkler. The estimated size of the so-called Pinwheel Nebula is about 300 AU (*credit Keck Observatory* via Wikimedia Commons) *Right*—**WR 140** in the constellation of Cygnus features concentric shells of dusty material that were expelled by another close binary star system, this time in episodic outbursts. Shock compression of the outflowing gas may have helped to generate the dust itself. The overall size of this remarkable system is estimated to be about 140,000 AU, or 2.2 ly (*credits* NASA, ESA, CSA, *JWST* MIRI & Ryan Lau et al.). **b** The O6.5-type star SAO 20,575 and the **Bubble Nebula** (NGC 7635) that it blew and is fluorescing. This nebula is located in the constellation of Cassiopeia at an estimated distance of 7100–11,000 ly. The ionized bubble spans 15 arcminutes on the sky, which translates to a linear size of 30–50 ly. This composite image was obtained with the *Hubble Space Telescope* through separate narrowband filters that passed [SII], H-alpha, and [OIII] line emission, respectively. The combination of sub-images was subsequently color-coded in red, green, and blue respectively. Many regions of twice-ionized oxygen are depicted in blue, indicative of high-energy photoionization and/or shock heating (*credits*NASA, ESA, *HST,* and the Hubble Heritage Team). **c WR 136** and the **Crescent Nebula** (NGC 6888) that it blew and is energizing. This more mottled nebula embodies another variation on the theme of powerful winds from massive stars impacting their surroundings. The Crescent Nebula is located in the constellation of Cygnus at a distance of 4100 ly. It spans about 18 arcminutes in the sky which translates to a linear size of about 21 ly—thus rivaling a classic HII region in breadth. This composite image was obtained at visible wavelengths through separate narrowband filters with the addition of X-rays from the *XMM-Newton* Observatory. The color coding presents X-rays in blue, [OIII] in green, and H-alpha in red. The yellow stars are artifacts of the color coding that was used. Here, the extensive X-ray and [OIII] emission indicate high-energy shock-heating by intense winds from the WR star (*credit* ESA/XMM-Newton, J. Toalá & D. Goldman with reference to Toalá et al. (2016) [https://arxiv.org/pdf/1512.01000.pdf]

Fig. 9.4 (continued)

This so-called supernova remnant, or SNR, expands at stunning speeds clocking at upwards of 10,000 km/s. The nebular gas sports a myriad of emission lines indicative of violent collisions underway. At visible wavelengths, emission lines from HII, [SII], [OII], [OIII], [ArIII], [NeIII], and various ions of iron up to [FeXIV] are often detected. Matters get even crazier at X-ray wavelengths, where emission lines from highly-ionized atoms announce their torrid circumstances. Suffusing the nebula, great swaths of energized plasma emit synchrotron radiation, as electrons traveling at near light speeds whirl around lines of magnetic force—losing energy in the process. And then there is the X-ray thermal continuum emission resulting from titanic shock waves heating the gas to coronal temperatures (see Fig. 9.5).

A good case in point is the remnant of **Supernova 1987A**, the nearest supernova to have exploded in the last 400 years. First seen as a naked-eye "star" in the Large Magellanic Cloud on February 24, 1987, the supernova shined with the light of more than 100 million Suns. The *Hubble Space Telescope* started observing it in earnest after its repair in December 1993.

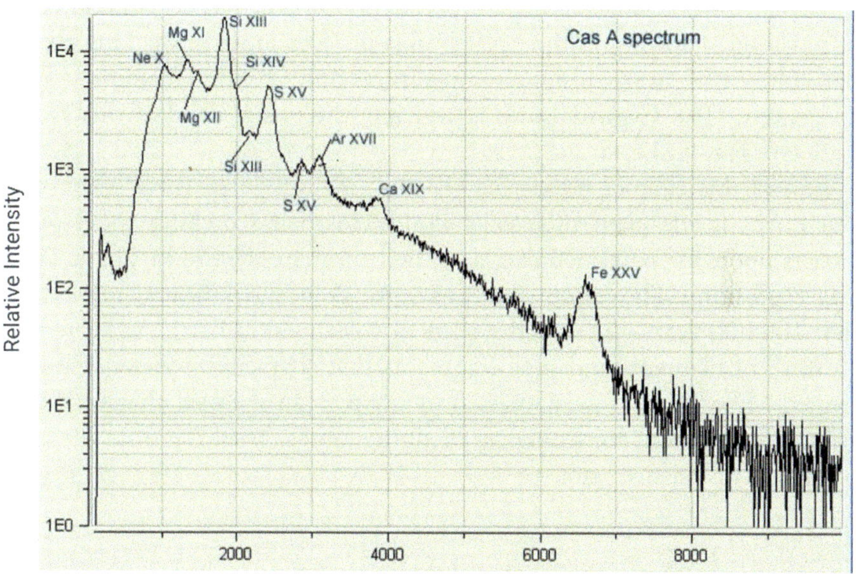

Fig. 9.5 X-ray spectrum of the **Cassiopeia A** supernova remnant shows numerous emission lines from highly-ionized atoms along with a sloping cascade of continuum emission. Both of these features are consistent with gas temperatures of 3–12 million Kelvins (*credit* Adapted from John Kolena, Duke University, https://webhome.phy.duke.edu/~kolena/snrspectra.html#casa)

The *HST*'s images showed a bright core ringed by a necklace of luminescent "beads." The necklace has been interpreted as a "light echo" from the 1987 outburst that illuminated material previously blown out by the former massive star's intense winds. In 2012 (25 years after the initial blast), the necklace markedly brightened, as collisions involving the supernova's ejecta shock-heated the annulus. The brightening has since subsided, while the central core has developed a fascinating substructure. Meanwhile, observations by *ALMA* have shown the emergence of a dust cloud that the SN's eruption likely generated. Here, we can bear witness to various gases having collisionally aggregated into microscopic grains of dust … one of several origins stories for the dust that pervades star-forming galaxies and makes rocky planets like Earth possible (see Fig. 9.6).

At first, a supernova remnant consists solely of material from the stellar explosion itself (as seen in SN1987A, Cassiopeia A, and the Crab Nebula). It expands at a rate dictated by energy conservation. Accordingly, both the nebular pressure and temperature decrease as the affected volume increases (see Fig. 9.7). But after the remnant has plowed up enough of its ambient interstellar surroundings, its expansion rate follows the physics of momentum conservation—slowing as the SNR builds up ever more mass. This phase is characterized by near-constant temperatures and prolific emission at many wavelengths. The **Cygnus Loop** is arguably the poster child of this evolved phase (see Fig. 9.8).

We end this chapter pondering the full life cycle of massive stars—from their births inside giant molecular clouds, to their scorching and scouring impacts upon the natal nebulae, and their ultimate demise in titanic supernova explosions. The latter eruptions leave their shock-heated marks for a scant 100,000 years—akin to the relics of volcanoes on Earth. What remains is a froth of filaments and shell fragments that characterize views of the far-infrared Milky Way, as we witnessed in Chap. 3. In the next chapter, we will explore the chemical consequences of all the foment resulting from the births and deaths of massive stars.

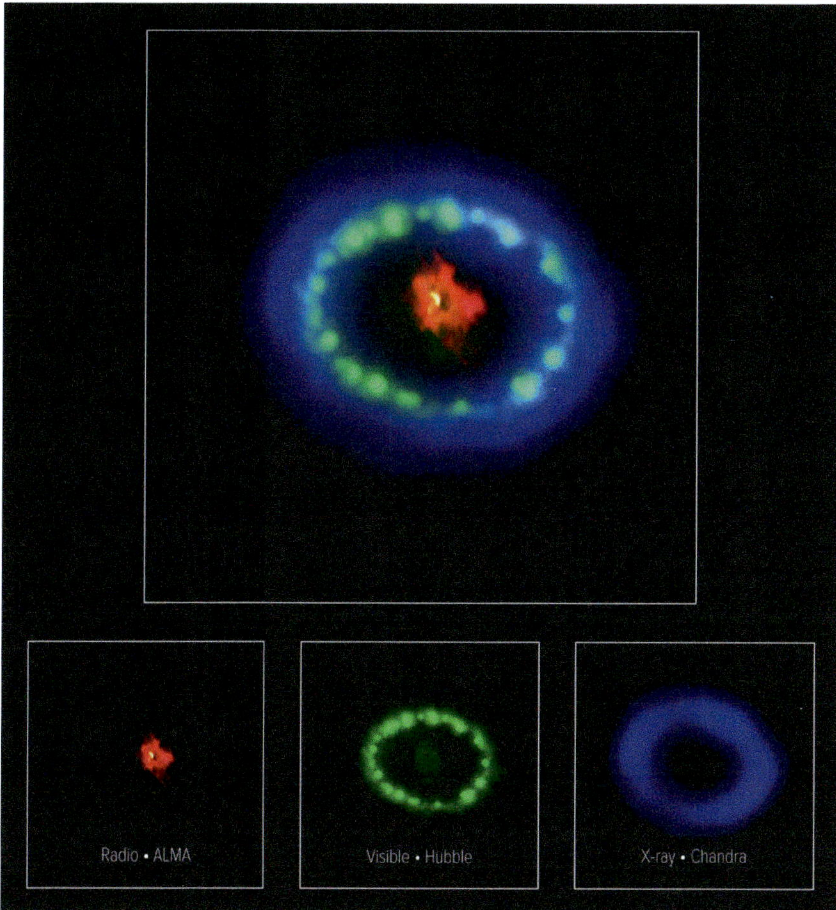

Fig. 9.6 The remnant of **Supernova 1987A** at millimeter, visible, and X-ray wavelengths. The mm-wave emission is from dust that has condensed out of the supernova debris. The visible emission shows a faint core with a bright ring of luminous beads that was blown by the former star's powerful winds and has since been energized by collisions with the outflowing debris. The ring is about 1 ly in diameter. The X-ray emission confirms that the hottest gas is concentrated in the ring (*credits* NASA, ESA, and A. Angelich [NRAO/AUI/NSF]; *Hubble*—NASA, ESA, and R. Kirshner [Harvard-Smithsonian Center for Astrophysics and Gordon and Betty Moore Foundation] *Chandra*—NASA/CXC/Penn State/K. Frank et al.; *ALMA*—ALMA [ESO/NAOJ/NRAO] and R. Indebetouw [NRAO/AUI/NSF])

Fig. 9.7 a **Cassiopeia A** is the remnant of a core-collapse supernova that exploded 350 years before the observations that produced the image shown here. At a distance of 11,000 ly, the observed remnant has a diameter of 18 ly. This composite image encodes the X-ray emission from million-degree gas in blue and green, visible emission from 10,000° gas in gold, and infrared emission from newborn warm dust in red. All three phases manifest the effects of shock waves rippling through the remnant and jets shooting outward. Much of the emission is from elements that were cooked up in the supernova blast itself. The remnant neutron star is the tiny blue dot near the remnant's center (*credits* X-ray - *Chandra*, visible - *HST*, and infrared - *Spitzer* observatories, Courtesy NASA/JPL-Caltech). **b** Two views of the **Crab Nebula** (M1), the remnant from a Type II supernova that was observed by Chinese astronomers and Anasazi Indians in 1054 CE, roughly 960 years before the imaging portrayed here. Located in the constellation of Taurus at a distance of 6500 ly, the elongated remnant spans about 13 ly. The *left panel* shows visible emission as imaged by the *Hubble Space Telescope.* Myriad tendrils highlight the shocking and shredding of ionized gas in the remnant. The central haze comes from synchrotron emission produced by electrons whizzing around magnetic field lines at near-light speeds. The *right panel* is a color-coded composite of infrared emission in red, visible light emission in green, and X-ray emission in blue. The bell-shaped X-ray emission delineates the torrid domain produced by powerful winds from the central pulsar (the central dot) (*credits Left* - NASA, ESA, J. Hester and A. Loll [Arizona State University]; *Right* - NASA, ESA, CXC, JPL-Caltech, J. Hester and A. Loll [Arizona State Univ.], R. Gehrz [Univ. Minn.], and STScI)

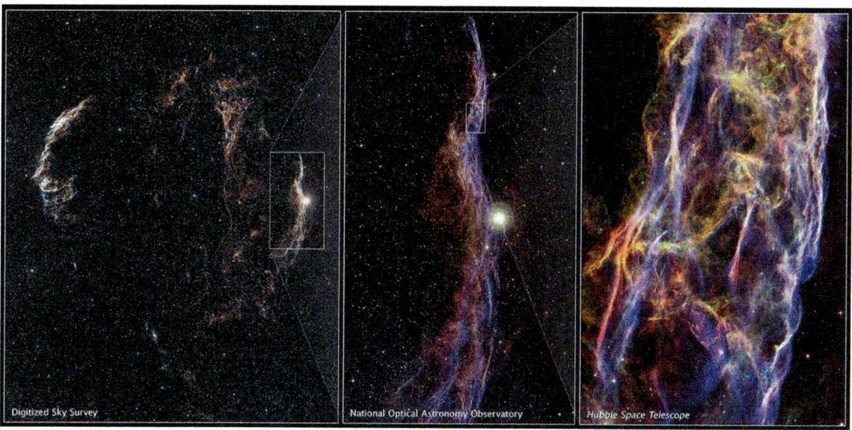

Fig. 9.8 The **Cygnus Loop** represents a more evolved supernova remnant, where much of the glowing material consists of the ambient interstellar medium that has been plowed up by the supernova's shock waves. This SNR is a popular target for amateur astronomers equipped with filters that pass the emission lines of H-alpha, H-beta, [NII], and especially [OIII]. At an estimated distance of 2400 ly, the entire loop has a diameter of 120 ly and an expansion age of 20,000 yrs. Like other supernova remnants, the nebulosity exhibits fine structure evocative of strong compressions by shock waves (*credits* NASA, ESA, the Hubble Heritage Team [STScI/AURA], Digitized Sky Survey [DSS, STScI/AURA, Palomar/Caltech, and UKSTU/AAO], and T. A. Rector [University of Alaska, Anchorage], and *WIYN*/NOAO/AURA/NSF)

10

Chemical Cocktails—Shaken and Stirred

Once a cache of massive stars emerges from its nebular cocoon, how do the hot stars' energizing presence affect the chemical processes taking place within the remaining galactic ecosystem? Alas, I wish I knew. The challenges to identify chemical compounds from the "forest" of emission lines in infrared and radio spectra are daunting. And the myriad chemical agents and pathways for creating molecules of ever greater complexity are even more bewildering (see Fig. 7.3). But I remain optimistic that astronomers will eventually sort out the possibilities.[1] In this spirit, I will bravely (naively?) forge ahead with an initial chemical examination of galactic ecosystems—from small molecular clouds to increasingly larger HII regions and starbursts.

The Corona Australis Molecular Cloud

Situated 20° below the Galactic Plane in the constellation of the Southern Crown, the Corona Australis Molecular Cloud can be readily spotted in spatially-filtered images of far-infrared emission (see Fig. 3.5). Its cometary appearance belies external agents having impacted the cloud. Parallax observations of member stars by the *Gaia Astrometric Observatory* indicate a distance

[1] Machine learning (aka Artificial Intelligence [AI]) may prove pivotal to this astrochemical endeavor, as it can drastically speed-up the identification of molecules from the forests of detected emission-lines. AI can also effectively rank the many chemical pathways that are available within any particular galactic ecosystem.

W. H. Waller, *Crucibles of Creation*, Astronomers' Universe,
https://doi.org/10.1007/978-3-032-17258-7_10

of only 490 ly, making it one of the nearest molecular clouds to us (see Fig. 4.5). Closer views reveal the B5-type variable star R CrA occupying and illuminating the densest portion of the cloud. R CrA is but one of several young variable stars making up the Coronet cluster at the "head" of the cloud (see Fig. 10.1). The largest members in the region, R, S, T, TY and VV Coronae Australis, are all ejecting jets of material which shock their surroundings and so produce Herbig-Haro objects (see Fig. 10.2).

Closeup views reveal fascinating structure in the reflection nebulae, where blue light from hot young stars has preferentially scattered off microscopic grains of dust (see Fig. 10.2). Here on Earth, similar behavior occurs, as sunlight scatters off air particles to produce the blue-sky phenomenon. The well-combed structuring in NGC 6726/6727 suggests the presence of ordered magnetic fields, while the cavitated morphology of IC 4812 could indicate prior stellar outflows and/or UV irradiation having sculpted the nebula. The near lack of red emission from ionized hydrogen tells us that this region is not home to the hottest O-type stars. Either they expired long ago, or they were never made. The relatively small size and low mass of this molecular cloud would suggest that it does not have enough material to forge such rare superstars. That said, the cloud plays host to many HH objects, thus confirming the presence of several young stellar objects of lesser mass undergoing intense outflows.

Fig. 10.1 The **Corona Australis Molecular Cloud** at visible wavelengths shows tendrils of obscuring dust that culminate in a cluster of young stellar objects at the cloud's densest "head." The cloud spans about 5° in the southern sky, which at a distance of 490 ly translates to a physical extent of 43 ly. The cloud contains about 7000 suns worth of molecular gas. The globular cluster NGC 6723 to the upper right of the imaged field does not belong to the cloud, lying much farther away at a distance of 28,000 ly (*credit* Courtesy of Fabian Neyer via Astronomy Picture of the Day)

Fig. 10.2 This annotated closeup of the Corona Australis Molecular Cloud's "head" shows multiple young blue stars enveloped by complex reflection nebulosity. HH refers to Herbig Haro objects—gaseous regions that have been shocked by energetic jets coming from the youngest stellar objects (*credit* Courtesy of Fabian Neyer via Astronomy Picture of the Day)

Spectroscopic observations by the *APEX* telescope and other mm-wave radio telescopes have provided evidence for molecules of diatomic hydrogen (H_2), carbon monoxide (CO), carbon sulfide (CS), hydrogen cyanide (HCN), the ethynyl radical (C_2H), formaldehyde (H_2CO), methanol (CH_3OH), and the positively charged molecular cation diazenylium (N_2H^+). Identification of this latter molecule suggests the presence of some partially ionized gas, perhaps due to UV irradiation by the young B-type stars in this formative region. Perhaps we are seeing here the first hints of chemical alteration by hot young stars.

Support for this interpretation comes from imaging at mid-infrared wavelengths by the *Spitzer Space Telescope* (see Fig. 10.3). Here, 24-micron emission from star-warmed grains of dust as well as 8-micron emission from polycyclic aromatic hydrocarbons characterize the infrared tableau. The PAHs, in particular, demonstrate that the young stellar environments in this relatively modest nebula still have what it takes to foster and energize complex organic molecules.

Fig. 10.3 Closeup of the **Coronet Cluster** in the "head" of the Corona Australis Molecular Cloud, as observed at mid-infrared and X-ray wavelengths. The total field of view here spans about 2 ly. The hot young stars in this region are warming ambient clouds of dust (shown in red) while exciting a filigree of PAH molecules (shown in green). The X-ray emission (shown in blue) traces young stellar objects along with a diffuse component of still unknown origin (*credits* X-ray: NASA, Chandra X-ray Center, Center for Astrophysics, and J. Forbrich et al.; Infrared: NASA/Spitzer Science Center, Center for Astrophysics, and IRAC GTO Team)

The Corona Australis Cloud, like the Taurus Molecular Cloud, should contain a rich assortment of complex organic molecules radiating at centimeter wavelengths. But that has yet to be confirmed. While observations at micron and millimeter wavelengths have revealed a tantalizing variety of molecules discussed here, spectroscopic observations at centimeter wavelengths remain mostly missing. Perhaps that is due to the cloud's southern equatorial latitude of – 40° which prevents sensitive cm-wave telescopes in the northern hemisphere such as the *Green Bank Telescope* from accessing it.

Until a large and suitably equipped single-dish radio telescope such as the *GBT* gets deployed for detailed spectroscopy in the southern hemisphere, we can only speculate on the complete molecular makeup of the Corona Australis Cloud.

The Rho Ophiuchi Molecular Cloud

Looking toward the Galactic Center, we can readily find the Rho Ophiuchi Molecular Cloud at a positive galactic latitude of about 17°. Its location provides a striking counterpoint to the Corona Australis Molecular Cloud which appears at a galactic latitude of negative 17° (see Fig. 3.5). Unlike the CrA cloud, the Rho Ophiuchi cloud has an equatorial latitude of – 24 ° and so can be accessed by many telescopes situated in the northern hemisphere. The resulting view at visible wavelengths is nothing short of stunning (see Fig. 10.4).

At a distance of about 460 ly, the nebular complex spans approximately 43 ly. Within this complex, many stellar and nebular dramas are unfolding. This nexus of star-forming activity was first revealed in the 1920s by E. E. Barnard from his pioneering photography of the night sky (see Fig. 4.1). Today, we can recognize that the nebular complex hosts long "rivers" of dusty molecular gas (see Fig. 3.5) that have spawned hundreds of young stellar objects. Several of these YSOs have illuminated the entrained gas and dust, thus producing the reflection nebulosity that is so evident in visible-light images. One YSO, Sigma Ophiuchi, has gone so far as to photo-ionize its surroundings and so produce an HII region.

Radio spectroscopy of the cloud has revealed the presence of several reactive molecules, including the hydroxyl radical (OH), carbon monoxide (CO), sulfur monoxide (SO), the cyanide anion (CN^-), carbon monosulfide (CS), the methylidyne radical (CH), nitroxyl (HNO), hydrogen cyanide (HCN), the formyl radical (HCO), the deuterated formyl cation (DCO^+), the diazenylium cation (N_2H^+), water (H_2O), ammonia (NH_3), formaldehyde (H_2CO), methanol (CH_3OH), the cyclic propenylidene (c-C_3H_2), and the linear (aliphatic) cyanodiacetylene (HC_5N). Many of these molecular detections are associated with dense cloud cores and YSOs. Some are consistent with moderately warmer temperatures near the YSOs.

Further insights on the dynamical and chemical character of the Rho Ophiuchi Molecular Cloud can be obtained at mid-infrared wavelengths, as shown in Fig. 10.5. Here, imaging by the *Spitzer Space Telescope* shows a rich mix of YSOs amid star-warmed dust radiating at 24-microns (coded red)

Fig. 10.4 The **Rho Ophiuchi Molecular Cloud** at visible wavelengths as observed from Mount Lemmon Observatory near Tucson, AZ. This complex of obscuring dust lanes, bluish reflection nebulosity, and roseate regions of ionized hydrogen (HII) gas delights the senses as much as it confounds full scientific understanding. To the lower left, the red supergiant Antares shines through a haze of reflection nebulosity, whose yellowish color parrots the M1-type star's ochre radiation field. To the lower right, the globular cluster Messier 4 has photobombed the natal scene, itself being located more than 7000 ly farther away. To the middle right, the B1-type binary star system Sigma Ophiuchi has created Sharpless 2–9, a classic HII region. To the top (north) of the field, the bluish reflection nebula IC 4604 scatters light from the B2-type binary star system Rho Ophiuchi AB (*credits* Courtesy of Adam Block/Steward Observatory/ University of Arizona—https://www.adamblockphotos.com/antares-and-rho-ophiuchi.html via CC BY-SA 4.0)

Fig. 10.5 Mid-infrared view of the star-forming activity within the Rho Ophiuchi Molecular Cloud, focusing on the dark nebulosity (Lynds 1688) to the left of the blue reflection nebula (IC 4604) as seen in the upper part of the prior visible-light image. The field of view spans 2.1° by 1.0°, which at a distance of 460 ly translates to physical dimensions of 17 ly by 8 ly. Myriad dust-embedded young stellar objects are evident at 3.6-micron wavelength (coded blue and magenta), while PAH molecules emitting at 7.7-microns (coded green) and larger grains of dust glowing at 24-microns (coded red) suffuse the nebular nursery. The plume-like nebulosity near the image's center surrounds the newborn star S1 Ophiuchi (*credits* NASA/JPL-Caltech, L. Allen [Harvard-Smithsonian Center for Astrophysics] and D. Padgett [Spitzer Science Center—Caltech])

along with PAH molecules radiating at 7.7-microns (coded green). Many of the YSOs are no more than 100,000 years old. The emitting PAHs appear widely distributed, as if energized by a multiplicity of sources.

More recently, the *JWST* has enabled us to look even closer at the seat of the action (see Fig. 10.6). The newborn Herbig Be star S1 is seen to have blown out and illuminated a large cavity in its maternal cloud, while also exciting the surface of this cavity to glow in the light of excited PAHs. Meanwhile, shocked molecular hydrogen (coded red) from multiple stellar outflows punctuate the tumultuous scene.

The Orion Nebula Bar—A Classic Photo-Dissociation Region (PDR)

To get a better handle on the effects of stellar UV irradiation on their natal galactic ecosystems, it is necessary to examine those thin transition regions, where cold molecules are dissociated into atoms, and then ionized into a plasma. These special working surfaces are known as photo-dissociation

Fig. 10.6 Closeup of near-infrared emission highlighting the nebular feedback produced by newborn stars in the Rho Ophiuchi Molecular Cloud. This high-resolution image by the *James Webb Space Telescope* measures only 0.8 ly by 0.8 ly. It reveals S1 Ophiuchi—a Herbig Be-type young stellar object that has excavated a cavity in the cloud that birthed it and is currently exciting the cavity's surface to glow in scattered light and in the light of PAHs at a wavelength of 3.3-microns (coded green). Multiple bipolar outflows from YSOs extend across the field, their emission from shocked molecular hydrogen at a wavelength of 4.7-microns is coded red here (*credits JWST*, NASA, ESA, CSA, STScI, Klaus Pontoppidan [STScI])

regions, or PDRs. As the nearest site of burgeoning massive star formation, the Orion Nebula (M42) provides the ideal site for exploring the PDR phenomenon (see Fig. 8.4). This HII region is especially well configured, as one of its PDRs has presented itself edge-on so that it can be viewed in cross-section (see Fig. 10.7).

Fig. 10.7 Visible and near-infrared views of the **Orion Nebula Bar**. Here, the working surface of the photoionized nebula has curled in our direction, thus revealing the photodissociation region (PDR) in cross-section. The ionizing Trapezium cluster of stars is located to the upper right, just beyond the imaged fields of view. At left, the visible-light image from *HST* depicts emission from ionized hydrogen (HII) in green and ionized sulfur ([SII]) in red. The [SII] emission, in particular, traces the ionization shock front that is eating into the molecular cloud. The ionization front is shot through with jets and HH objects from YSOs that are otherwise hidden from view. At right, the near-infrared image from *JWST* depicts reflection nebulosity and embedded stars in blue, PAH emission at 3.3-micron wavelength in green and red, and emission from star-warmed dust in red. By comparing the two images, an offset in the bar's location can be discerned. Just compare the respective bars' positions relative to the bright star Theta-2 Orionis A in the lower middle of the field, where the ionized bar is situated significantly farther from the star (*credits Left*—NASA/STScI/Rice Univ./C. O'Dell et al.—Program ID: PRC95-45a; *Right*—NASA, ESA, CSA, Data reduction and analysis: PDRs4All ERS Team; graphical processing S. Fuenmayor & O. Berné)

The offset across the bar between the molecular, atomic, and ionized components amounts to only 0.3 ly, less than 2% of the nebula's overall expanse. Within this thin skin, all sorts of photo-chemical transformations are taking place. The state of hydrogen shifts from ionized H^+ above the ionization front, to atomic HI and molecular H_2 within the bar. Similarly, carbon shifts from ionized C^+ at the ionization front, to atomic C and molecular CO deeper into the molecular cloud. Other hydrocarbon molecules mingle with CO in the more shaded part of the bar. These include the formyl cation (HCO^+), ethynyl (C_2H), acetylene (C_2H_2) the radical propnylidyne (C_3H) and its positively charged cation (C_3H^+), the cyclic cyclopropenylidene (c-C_3H_2), and the linear radical butadiynyl (C_4H_2). All of these molecules are implicated as part of the rich chemical reactions resulting from the UV irradiation by the hot stars in the Trapezium cluster.

Within the Bar, a wide variety of PAHs announce their presence at near- and mid-infrared wavelengths (see Fig. 10.8). The sundry PAHs arrange themselves in layers that depend on the intensity of UV irradiation. The larger PAHs are found closer to the exposed ionization front and are often ionized, while the smaller more fragile PAHs reside deeper within the Bar and molecular cloud, where the UV radiation field is less intense. Although the chemical pathways among the various small and large molecules within PDRs have yet to be sorted out, we can appreciate that the energizing influence of UV starlight from massive stars can yield caches of new organic species, some of which could be biogenic.

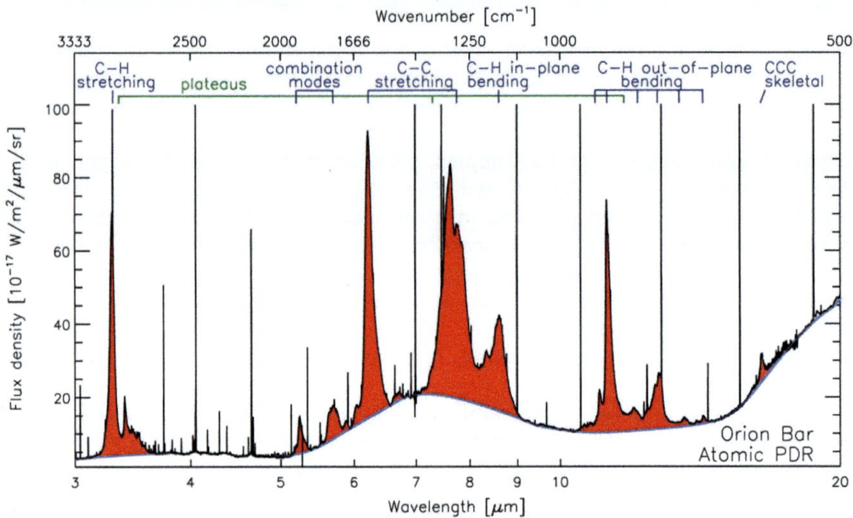

Fig. 10.8 A composite infrared spectrum of the Orion Bar as obtained by the *JWST*. An assortment of emission bands from PAHs in various states is highlighted in red. They include the characteristic bands at 3.3, 6.2, 7.7, 8.6, 11.2, and 12.7 micron wavelengths. The precise wavelengths, relative intensities, and bandwidths are all used to diagnose the types of PAHs that are present (*credits* R. Chown et al. (2024), *Astronomy & Astrophysics,* vol. 65, p. A75 [https://doi.org/10.1051/0004-6361/202346662])

The Heart and Soul Nebulae—Twin Giant HII Regions

A favorite of astrophotographers, the Heart and Soul Nebula (Sharpless 2–190 and 2–199)[2] blaze forth within the Perseus spiral arm—more than 6000 ly away. Together, they embody galactic ecosystems that are fully involved in starburst activity. Herein, star clusters containing dozens of massive ionizing stars have photo-excavated enormous cavities in their once molecular wombs (see Fig. 10.9). These cavities dwarf the one in the Orion Nebula by more than an order of magnitude, making them likely products of additional expansive dynamics from intense stellar winds and supernova blasts. Closer examination of the Soul Nebula has revealed multiple generations of newborn stars—with the youngest distributed along the cavity walls and the oldest well inside (see Fig. 10.10). Here, we find compelling evidence for star-formation having propagated outward. In this scenario, the UV starlight, winds, and supernovae from earlier stellar generations have compressed the ambient gas into new star-forming regions.

Throughout these nebulae, vast realms of emitting PAHs and dust trace the zones where organic chemistry is being activated by the resident hot stars. What new molecules await discovery? Radio CO and OH surveys indicate close to 100,000 suns worth of molecular gas in these regions, mostly in the form of diatomic H_2. But the necessary wideband spectral-line surveys of the molecular emission at cm-, mm-, and sub-mm wavelengths have yet to reap a complete dossier of molecules. One survey of the radio-bright source W3 (to the right of the Heart nebula in Fig. 10.9) by the *Submillimeter Array (SMA)* in Hawaii found molecules of carbon monosulfide (CS), sulfur monoxide (SO), sulfur dioxide (SO_2), hydrogen cyanide (HCN), thioformaldehyde (H_2CS), isocyanic acid (HNCO), and methanol (CH_3OH)—each with fascinating spatial variations. Note the rich sulfur chemistry in this one small region. Similar sulfur-bearing molecules can be found in the atmospheres of some giant exoplanets. Meanwhile, other parts of the larger nebular complex remain unassayed. Any takers?

[2] Sh 2-190 and Sh 2-199 have other identifications, including the radio sources W3, W4, and W5 as well as the stellar clusters IC 1805 and IC 1848.

Fig. 10.9 The **Heart and Soul Nebulae** in the constellation of Cassiopeia, as imaged by the *Wide-field Infrared Survey Explorer (WISE).* At a distance of about 6,000 ly, the 5.5 by 3.9°. field of view spans 580 by 410 ly. The Heart Nebula is seen to the right, with its resemblance to the chambers in a human heart. The Soul Nebula to the left also features large cavities that were excavated by the intense UV radiation and strong winds from newborn massive stars. This infrared image has stars emitting at 4-micron wavelength depicted in blue, PAHs emitting at 12-microns depicted in green, and star-warmed dust emitting at 22 microns depicted in red (*credits* NASA/JPL-Caltech/UCLA, *WISE*)

The Starburst Galaxy—M82 (NGC 3034)

In this chapter, we have explored the physical and chemical character of galactic ecosystems over a wide range of sizes and masses. We have learned that even in the smallest and least substantial GEs, jets from young stellar objects can be seen injecting kinetic energy into their nebular birthsites. The larger GEs play host to proportionately larger clusters of stars which, in turn, include several massive ionizing stars. These stars' intense UV radiation, strong winds, and ultimate supernova explosions carve out vast cavities in their birth clouds—with the working surfaces betraying ionized, atomic, and molecular phases of energized gas. Amidst all of these GEs, complex organic molecules have been identified, some of which have biogenic properties. Although the relationship between a GE's powering and its suite of molecules has yet to be worked out, there appear to be subtle hints that the

Fig. 10.10 Closeup of the **Soul Nebula**, as imaged in the infrared by the *Spitzer Space Telescope*. The field of view covers 280 by 190 ly. The nebula features resistive pillars of gas and dust protruding inward from the cavity walls. Protostars are evident at the tips of some of these ablating pillars. Similar to the *WISE* image, stars imaged at 3.6-micron wavelength are depicted in blue, PAHs imaged at 8-microns are depicted in green, and star-warmed dust grains imaged at 24-microns are depicted in red (*credits* NASA, JPL-Caltech, *Spitzer Space Telescope,* Harvard-Smithsonian Center for Astrophysics, Lori Allen and Xavier Koenig)

more powerful GEs contain a greater proportion of ionized molecules as well as more energized PAHs.

We end this chapter considering the upper end of GE powering, namely the Cigar Galaxy (aka M82, NGC 3034)—one of the nearest galaxies to host system-wide starburst activity (see Fig. 10.11). Located 12 Mly away in the constellation of Ursa Major, M82 appears as a disk galaxy that is situated nearly edge-on to our line of sight. It was classified as an amorphous or irregular galaxy for many decades, but more recent near-infrared imaging has revealed two spiral arms in the inclined disk, thus suggesting some sort of spiral galaxy classification. Its estimated diameter of 40,000 ly is about 40% that of the Milky Way galaxy, making it one of the smaller galaxies to host spiral structure. Yet, that has not prevented M82 from putting on quite the show.

Extending away from the galaxy's core, two bipolar plumes of gas and dust fan outward for thousands of light-years. Astronomers recognize these plumes as tracing superwinds that are being driven by multiple supernova explosions

Fig. 10.11 Composite image of M82 depicting infrared emission from PAH molecules in red, visible emission from ionized hydrogen in orange, visible starlight in green, and X-ray emission in blue. These components were respectively obtained by the *Spitzer, Hubble,* and *Chandra* space telescopes. The angular field of view is 7.9 arcmin across which translates to 27,000 ly at the galaxy's distance of 12 Mly (*credits* NASA/JPL-Caltech/STScI/CXC/UofA/ESA/AURA/JHU)

near the galaxy's core. These winds have been clocked at several thousand km/s, greatly exceeding the velocity of escape from this galaxy. Such outflows are largely responsible for depleting the galaxy's star-forming gas while feeding the intergalactic medium.

Back to the core, one can find massive clusters of newborn stars arrayed in a conga line. The current rate of star formation in this relatively small region is breathtaking—more than ten times that occurring in the entire Milky Way galaxy. Such profligate starbirth activity cannot last for long, as the available gas for forging new stars would limit the birthing time to no more than 200 million years. Given these drastic circumstances, the term "starburst galaxy" is fully justified.

From multi-wavelength studies, astronomers have found the bipolar outflows to manifest granular, molecular, atomic, ionized, and coronal phases

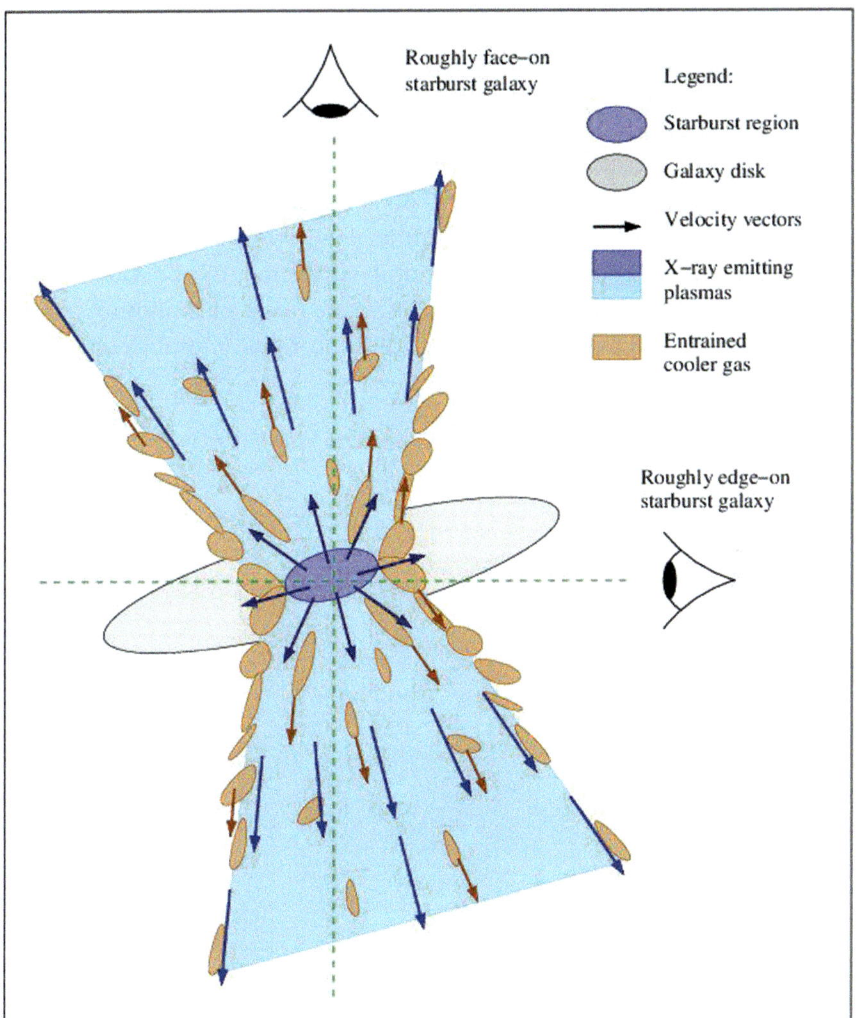

Fig. 10.12 Schematic of a multi-phase superwind from a starburst galaxy such as M82. We see M82's disk nearly edge-on. The X-ray plasma at millions of Kelvins fills the two opposing outflows. The cooler ionized and molecular gas clouds are entrained within the hot coronal plasma (*credits* Adapted with permission from D. K. Strickland, A. Hornscheimer, et al. 2009)

at temperatures of tens, hundreds, thousands, and millions of Kelvins, respectively. The hottest phases exert the greatest pressures, inducing the cooler phases to go for the ride (see Fig. 10.12). Within the starbursting core, a witch's brew of molecules have been identified at radio wavelengths, including neutral and ionized carbon dioxide (CO), carbon monosulfide (CS), carbon mononitride (CN), nitrogen monoxide (NO), the formyl cation

(HCO^+), the isoformyl cation (HOC^+), the hydronium cation (H_3O^+), hydrogen sulfide (H_2S), the cyclic hydroxyl (c-C_3OH), formaldehyde (H_2CO), thioformaldehyde (H_2CS), hydrogen cyanide (HCN), cyanamide (NH_2CN), formamide (CH_3ON), methanol (CH_3OH), and methylacetylene (CH_3CCH). Such a rich mix of molecular species suggests the presence of dense molecular clouds in a bath of energizing photons from massive newborn stars. Could life take hold on a planet within such a fierce environment? Complex molecules are certainly present, but the UV radiation fields might be too extreme for the delicate task of building biogenic macromolecules akin to carbohydrates, lipids, proteins, and nucleotides—the subject of the next (and last) section.

Part IV

Galactic Ecosystems and the Origins of Life

So many atoms, clashing together in so many ways,
as they are swept along through infinite time by their own weight,
have come together in every possible way and realized everything
that could be formed by their combinations.

—Titus Lucretius Caras in *On the Nature of Things* (55 BCE)

11

Forging the Elements of Life

On Earth, there are six key elements that underlie all forms of life. They are carbon (C), hydrogen (H), nitrogen (N), oxygen (O), phosphorus (P), and sulfur (S). Together they spell CHNOPS which sounds a bit like "Schnapps" and so is rather memorable in an amusing sort of way. How they rose to the top of the biogenic heap remains somewhat puzzling, as they do not represent the most abundant elements in the cosmos, on Earth, or even in our oceans (see Fig. 11.1).

Some of these disparities are self-evident. Were silicon (Si) as abundant in humans as in the Earth's crust, we would be way too rocky. Similarly, were sodium (Na) and chlorine (Cl) as abundant in humans as in the oceans, we would be too salty causing imbalances in osmotic pressures at the cellular level. Given these anomalies in elemental abundances, how can we explain the salience of CHNOPS in humans and other lifeforms? The answer comes down to examining the chemical structures of the four key macromolecules that are present in every living cell—namely carbohydrates, lipids, proteins, and nucleotides. Carbohydrates provide structural support and fuel every cell's metabolism. Lipids make up every cell membrane and wall among other functions. Proteins provide a dizzying number of services—from building and repairing tissues to catalyzing biochemical reactions and doing much of the work within cells. Nucleotides encode the reproductive process and the construction of proteins. Here, we can see the CHNOPS squad playing key roles as part of the macromolecules (see Fig. 11.2).

W. H. Waller, *Crucibles of Creation*, Astronomers' Universe,
https://doi.org/10.1007/978-3-032-17258-7_11

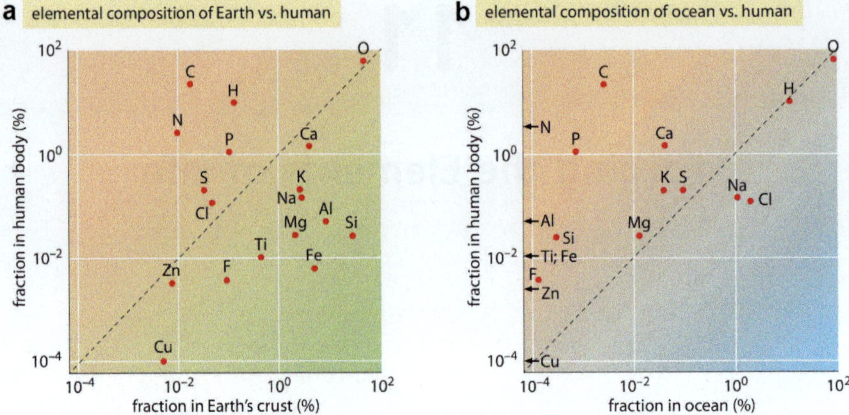

Fig. 11.1 Fractional abundances of the elements in (**a**) the Earth's crust vs. the human body, and in (**b**) the ocean versus the human body. In both comparisons, the elements of C, N, P, and S are drastically overabundant in humans. In the comparison with Earth's crust, only O is comparable in humans, while in the comparison with the ocean, both O and H are comparable in humans—the consequence of our watery composition (*credit* From "Cell Biology by the Numbers" by Ron Milo and Rob Phillips, with permission. [https://book.bionumbers.org/what-is-the-elemental-composition-of-a-cell/])

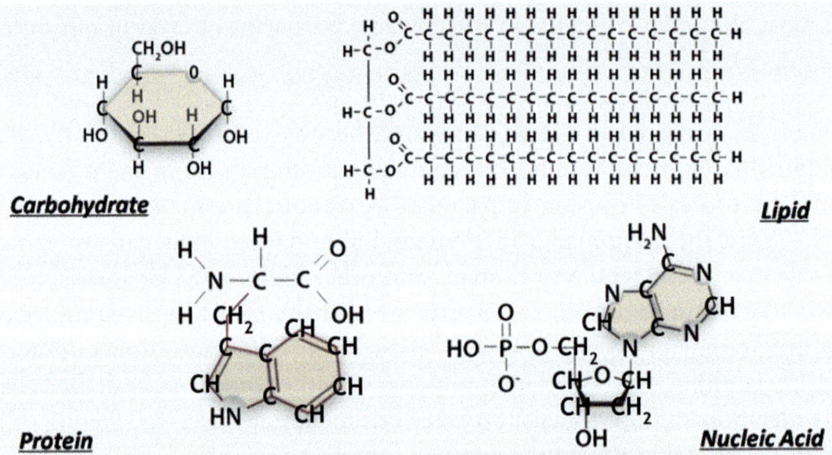

Fig. 11.2 Chemical structures of the four macromolecules essential to all life on Earth. Both carbohydrates and lipids are made solely of carbon, hydrogen, and oxygen. The cyclic (ring) molecules have corners that are occupied by a variety of carbon, hydrogen, and other atoms. Proteins are built from amino acids which, in turn, contain nitrogen as well as carbon, hydrogen, and oxygen. Nucleic acids are made from nucleotides which contain phosphorus and nitrogen in key positions (*credit* Adapted with modifications from iteachly.com at https://iteachly.com/macromolecules-biology-activity/)

Cosmic Alchemy—The Diverse Origins of Elements

While life on other planets could rely on other sorts of macromolecules, it is still worth investigating the CHNOPS set of essential elements on Earth in terms of their chemical capabilities and cosmic origins. Most of the hydrogen and helium atoms are known to have been created in the microseconds following the Big Bang. The other elements were all forged in the cores of stars, during supernova explosions, or in collisions between neutron stars—the compact remnants of massive stars (see Fig. 11.3). Carbon, along with other elements in its valence group, is equally facile in taking on or donating electrons in the bonding process. That, plus its greater cosmic abundance among other elements in its group, explains why it often forms major parts of complex molecules such as the macromolecules of Earthly life. Carbon is created in the normal sequence of thermonuclear reactions that occur in the thermonuclear cores of evolved giant stars. The sequence proceeds from hydrogen to helium and then to carbon. Stars of the Sun's mass will exhale these heavier elements in the form of planetary nebulae which then will diffuse into the ambient interstellar medium.

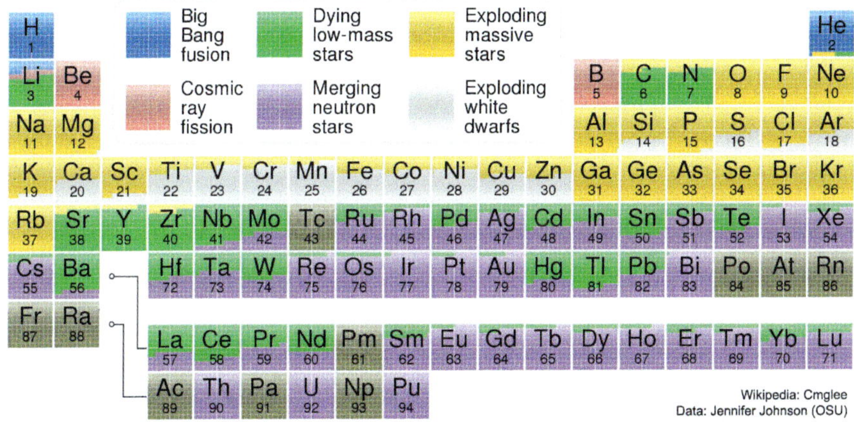

Fig. 11.3 Periodic table of the elements, with color-coding to indicate the cosmic object and process by which the element was made. The valence groups correspond to the vertical columns in this table, where the group number denotes the number of electrons in each atom's outer shell. For example, hydrogen is the lightest element in the first valence group, while carbon is the lightest element in the fourth valence group, and oxygen is the lightest element in the sixth valence group (*credit* Jennifer Johnson and Ohio State University via Wikipedia)

If the star is sufficiently massive, carbon will fuse into neon, oxygen, silicon, and iron. Oxygen is highly reactive (think of rust and fire) and so plays important roles in energizing organisms. Nitrogen is formed as part of the catalytic CNO reactions in massive stars. This element forms pivotal parts of nucleic acids, the amino acids that make proteins, as well as the chlorophyll molecules that enable plant photosynthesis. Sulfur and phosphorus are also thought to get made in massive stars. Sulfur is found in several amino acids and is used as a macronutrient by all living organisms. It has been found in the "black smokers" on the ocean floor, where bacteria respire it instead of oxygen. Phosphorus plays vital roles in cell membranes, nucleic acids, and the ATP molecule that energizes many biological functions. All these elements will be violently ejected, when the supergiant star explodes as a supernova. Many of the elements heavier than iron will be forged in the supernova itself, where there are lots of neutrons available for the rapid (r) and slow (s) processes of building up atomic nuclei.

The advent of gravitational wave astronomy began in 2015 with the detection of a distinctive "chirp" by the twin Laser Interferometer Gravitational-Wave Observatory (*LIGO*) facilities. This signal was interpreted as a disturbance in the fabric of spacetime that had propagated to us from two colliding stellar black holes. In 2017, another fainter but longer chirp was detected. This time, the physicists interpreted the detection event as resulting from the collision of two neutron stars. Such an intense collision would have liberated lots of neutrons which, again, could glom onto any pre-existing atomic nuclei to make even heavier nuclei. All of these alchemical mechanisms are elucidated in Fig. 11.3, where the color coding indicates which mechanism is most active.

Most of the heavy elements do not provide vital services to biological processes. Indeed, they are often regarded as poisons. However, some metals play fascinating roles as helpful electron transporters and as catalyzers of various reactions. These metals include sodium (Na), potassium (K), magnesium (Mg), and calcium (Ca) in the first two valence groups along with six of the transition metals—manganese (Mn), iron (Fe), cobalt (Co), copper (Cu), zinc (Zn) and molybdenum (Mo). Check the ingredients on your bottle of vitamins, and you will see some of these metals listed as "minerals." Like the playlist of essential elements for biotic processes, these elements were all formed in the thermonuclear guts of stars, during supernova explosions, or from the collisions of neutron stars (see Fig. 11.3). Given such a stellar pedigree, we can rightfully claim that we are "starstuff," as connected to the cosmos as the galactic ecosystem that spawned us.

12

Building Habitable Planets

What does it take for life to take hold on a planet? That is one heck of a loaded question. First, we don't really know what forms of life are possible. We don't even know what differentiates life from abiotic systems, e.g. rocks, ices, and various slimes that contain complex organic chemistries. This uncertainty has plagued the field of astrobiology since the 1960s, when the field was called exobiology. At the risk of being self-referential, we might consider the following functions that have characterized all life on Earth.

- The life form must be contained in some way, so that it can interact with its environment as an autonomous entity. On Earth, the containment unit is the cell, wherein all sorts of chemical processes take place that are unique to the cell itself.
- The life form must exchange matter and energy within itself and with its environment. We call this activity metabolism. Sometimes, we can recognize the manifestations of metabolism as animate respiring, feeding, excreting, circulating, moving, and growing. Other times the metabolic activity will be more subtle, as in biochemical signaling. All this electrochemical activity tends to put organisms in thermodynamic dis-equilibrium with their surroundings—another tell-tale sign of life.
- The life form must be able to reproduce. Otherwise, it must rely on tremendous longevity to achieve "success" as a living entity.
- As a corollary to reproduction, the life form must participate in the evolutionary process of variation and selection in response to its changing environment.

W. H. Waller, *Crucibles of Creation*, Astronomers' Universe,
https://doi.org/10.1007/978-3-032-17258-7_12

Here is a discussion from *The Astrobiology Primer* v. 2.0 that distills these various considerations into a more succinct description of life …

Currently, the most popular definition of life was conceived by a NASA effort to create a "working definition" for their Exobiology and Astrobiology research programs. The result, most often cited as Joyce (1994), defines life as "a self-sustaining chemical system capable of Darwinian evolution." A major strength of this simple and concise definition is that it distinguishes life by the evolutionary process rather than its chemical composition. The definition alludes to biological chemistry through the general term of "self-sustainment."

From my personal experiences with plants, pets, and parents—I would add that all living organisms must someday die. The mortality of all life makes the act of living all the more precious and worth understanding. We are now getting into the *meaning* of life, which I will have to leave to faith leaders, philosophers, and Monty Python's Flying Circus.

A second nagging uncertainty to the question of planetary habitability is that we don't even know whether planets represent the only locales for hosting life. In his book *Black Cloud*, full-time cosmologist and part-time science fiction writer Fred Hoyle imagined a vast interstellar cloud that acted as one rather nasty organism that later came to terms with its human interlocutors. It relied on signaling at radio wavelengths to carry out its various functions. Other sci-fi stories that feature nebulae in starring roles include six episodes of *Star Trek*, Episode 4 of *Star Wars* where a space slug (exogorth) is found in an asteroid field inside a nebula, and *Picard*, season 3, episode 4 ("The Cloud"), where the protagonists enter a nebula to tap into its power reserves but end up discovering that the cloud is a sentient organism.

But planets do have a lot going for them as possible hosts for life to flourish. First, they offer a gravitating surface for biochemicals to gather and concentrate. Second, they can retain an atmosphere which, in turn, can pressurize bodies of water against sublimating away. Third, their heat of formation and lingering radioactivity can provide helpful thermal energy for various biogenic reactions. Lastly, those planets with spinning liquid metal cores can generate magnetospheres that will protect any organisms on the planetary surface from cosmic rays and other harmful ionized particles coming from their mother star. So, let's get on with it and consider the options for building habitable planets.

Choosing a Suitable Star

Not all stars are amenable to hosting habitable planets. At the high-mass end, stars burn especially bright—thus exhausting their thermonuclear supplies in short order. Figure 12.1 highlights the drastic dependence of stellar lifetime on mass, whereby the lifetime declines as the 2.5 power of mass. Considering their briefest of lifetimes, the highest-mass O- and B-type stars would explode into the void well before the simplest organisms could have emerged on any Earth-like planet around them. If we want to allow sufficient time for the business of biochemistry to take hold, then we would do well to stick with stars of spectral types M, K, G, and F (and corresponding masses of 0.1–1.6 solar masses, see also Table F.2). That would leave you about 45 qualifying stars to explore within 15 ly of the Sun.

Then there is the issue of environmental stability. The lowest-mass M-type stars are noted for powerful flaring activity during the first 4 Gyr after their formation. This is because they are fast rotators at birth and have yet to spin

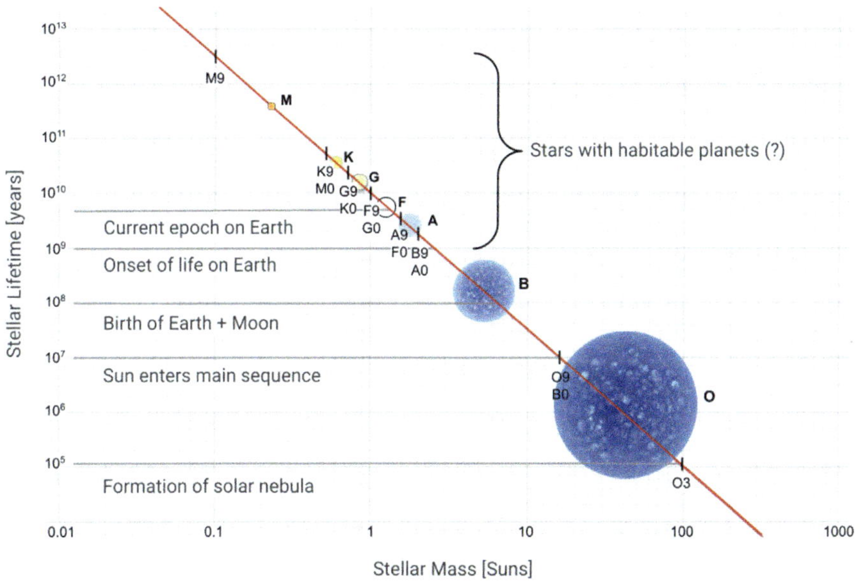

Fig. 12.1 Lifetimes of stars during their H-burning main-sequence phases as a function of stellar mass. The drastic decline of stellar lifetime with stellar mass appears as a straight line on this log–log graph. On Earth, it took a billion (10^9) years following its formation for any hint of microbial life to emerge. This sort of gestation time places a limit on the types of stars that can host habitable planets. A more detailed timeline that refers to the birth and evolution of the Solar System can be found at https://en.wikipedia.org/wiki/Formation_and_evolution_of_the_Solar_System#Timeline_of_Solar_System_evolution (*credits* W. H. Waller and L. H. Slingluff)

down into a more sedate state, where their churning magnetic fields no longer induce harmful ejections of matter and energy. Any M-type star with a Me classification has been identified as having spectral-line emission indicative of flaring. These bad boys account for 45% of all the M-type stars within 15 ly of the Solar System (where we have a complete accounting). Best beware! If I was to bet on a favorable star in terms of hosting habitable planets, I would stick with stars of K, F, and G-types. That would amount to ten such stars within 15 ly of the Sun. There would remain 20 unflaring M-type stars with which you can take your chances.

Delineating Habitable Zones

The typical definition of habitability is for liquid water to exist on the surface of the planet. The water molecule is an amazing substance whose strong electric polarity encourages various chemical hookups—and so provides a favorable solvent for biochemistry as we know it.[1] The water must be in liquid form, however, to promote these various reactions. We are most familiar with the reactions that involve oxygen as an energizer or byproduct, but other biotic reactions can ensue within a watery medium. Explorations of Earth's ocean bottom have shown that life can thrive in hot "black smokers," where sulfurous chemistry has replaced oxygenic chemistry in the biotic functions of the bacteria and the tube worms that feed on them. Methanogens are single-celled archaea that combine carbon dioxide and molecular hydrogen in water solution to produce methane. You can find them at work in rice fields and the guts of ruminants. So, we have a good case for liquid water on planetary surfaces promoting the business of life in many milieus.

That said, robotic explorations of the outer planets and their moons have shown that the liquid water need not hew to the planetary surfaces. Several of Jupiter's moons—including Europa, Ganymede, and Callisto—are thought to contain oceans beneath their icy crusts. The *Cassini* orbiter imaged Saturn's moon Enceladus spewing geysers of water. When it flew through the outflows, its mass spectrometer registered various hydrocarbons

[1] Other molecules could serve as liquid solvents, as long as the temperature is in the correct range. For example, the ammonia molecule (NH_3) has a significant electric polarity for encouraging chemically active mixtures. To be in liquid form, it can be no warmer than − 33 °C, unless it is in solution with some other solvent (e.g. water on Earth, as in our household cleansers). The surface of Saturn's moon, Titan, has temperatures of − 180 °C and so would qualify as a possible venue for hosting pools that contain liquid ammonia. We already know that Titan has a nitrogenous atmosphere and lakes of liquid methane (CH_4) which is non-polar. Perhaps ammonia is part of the mix there, helping to make new chemical connections—albeit at a cryogenic pace.

that await further assaying. What will future missions, beginning with the *Europa Clipper* scheduled to reach Europa in 2030, tell us about habitability within these liquid layers? If we allow for liquid water embedded within otherwise frozen worlds, our definition of a habitable zone grows beyond what we can readily delineate. So, let's first stick with the typical definition where surface water must be in liquid form. Given this criterion, the surface must have a temperature that ranges between 0 and 100 °C. Even this thermal requirement must be tempered by the atmospheric pressure at the planet's surface. Again, we are forced to presume pressures similar to those on Earth's surface to get anywhere.

Once we allow for these presumptions, we can go ahead and determine a star's habitable zone based on its luminosity. That involves determining the stellar flux as a function of distance from the star and then estimating the thermal response of the planet's surface at that distance. The latter depends on the surface's reflectivity, or albedo, with higher albedos yielding lower surface temperatures. Any greenhouse gases in the planet's atmosphere must be considered as well. The resulting distribution of habitable zones is shown in Fig. 12.2, where only main-sequence stars are considered. For hotter, more luminous main sequence stars, the habitable zones are more distant and extend over a greater range of radii, thus increasing the likelihood of an exoplanet occupying the zone. But it would be a fool's errand to rule out the cooler and fainter M-type stars, despite their propensity to flare, as they are so much more numerous.

What Have We Found?

Since the first exoplanetary detection in 1990, groundbased telescopes and spaceborne missions have racked up more than 5800 detections of exoplanets in orbit around more than 4300 stars. The measured radii and masses yield average densities consistent with the full gamut of interior properties—from gas balls to water worlds, rocky orbs, metal marbles, and hybrids like our own Earth (see Fig. 12.3). Any of these planet types could have locales where life could take hold, provided the right mix of chemicals can co-mingle within a favorable liquid solvent such as water. The water need not be a classic liquid, but rather a vapor that is not so hot as to sterilize the lifeforms.

If we restrict ourselves to rocky planets, there remain several dozen exoplanets that have been found to orbit within their stars' habitable zones. Indeed, the future looks bright for further investigating these planets in terms of their atmospheric compositions and surface properties. The *JWST* has

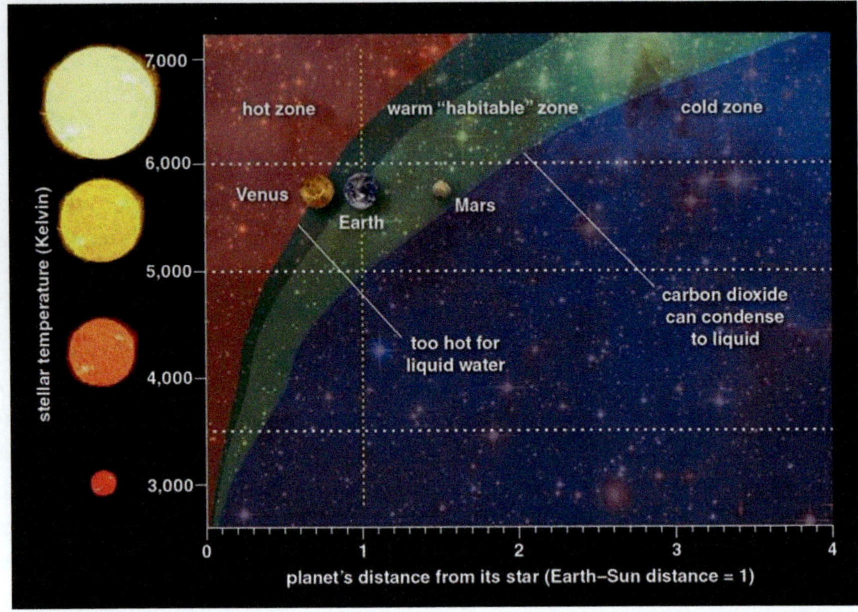

Fig. 12.2 Plot of habitable zones for a range of stellar temperatures with Venus, Earth, and Mars populating the graph at the Sun's surface temperature of 5800 K. Here, the plotting assumes the same mix of gases as are present in Earth's atmosphere. The inner limiting radius is set by the onset of liquid water vaporizing and thus producing intense greenhouse warming. The outer limiting radius is set by the onset of carbon dioxide condensing into liquid thus nixing its greenhouse warming. The ice-clad moons of Jupiter and Saturn would lie well beyond this limit, yet several of them contain liquid water beneath their icy surfaces (*credits* From Kevin Heng in *American Scientist,* May–June 2016, vol. 104, no. 3, p. 146, illustration by Barbara Aulicino, with permission. Data source: NASA, PHL@UPR, https://www.americanscientist.org/article/the-imprecise-search-for-extraterrestrial-habitability.)

begun these investigations, obtaining spectra of rocky exoplanets during transits in front of their stars. So far, the results have vexed the astronomers, as the expected atmospheric features in the spectra are mostly lost in the noise. To do better, astronomers are setting their sights on the dedicated *Habitable Worlds Observatory* that is currently at the planning stage, with an anticipated launch date in the 2040s.

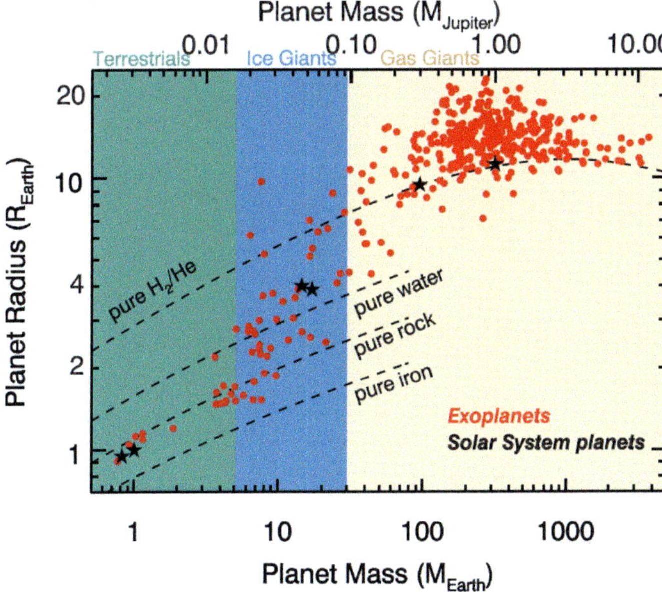

Fig. 12.3 A plotting of the exoplanetary masses and radii as detected and measured up to 2018. This sort of plotting enables scientists to distinguish populations of metallic, rocky, watery, and gaseous planets. For reference, the starred planets indicate (from left to right) Venus, Earth, Uranus, Neptune, Saturn, and Jupiter (*credit* From Exoplanet Science Strategy, National Academies Press (2018), https://nap.nationalacademies.org/read/25187/chapter/4#24)

13

Crafting Biomolecules

In 1952, Harold Urey and Stanley Miller at the University of Chicago conducted a laboratory experiment that simulated the physical conditions that were then thought to characterize the primordial Earth (see Fig. 13.1). A large flask was filled with an "atmosphere" of gases that included molecular hydrogen, water, methane, and ammonia. To provide the activation energy for any possible reactions in this medium, they introduced electrical discharges on a continuous basis. The flask of gases was also fed vapors from another flask of heated water to simulate the ocean. These vapors were then cooled and collected in a glass "trap" as condensation products. After a week of zapping the mix, they found that a red tar-like substance had coated the trap. Analysis of this residue revealed evidence of various organic compounds, including several different amino acids. For the first time in human history, they had synthesized biochemicals from scratch.

Since Miller and Urey's pioneering experiment, geochemists have questioned the atmospheric mix that they used. According to these contemporary scientists, early Earth likely hosted an atmosphere dominated by carbon dioxide and diatomic nitrogen rather than diatomic hydrogen, methane, and water. Modifying their experiments according to what they believed was extant in "Hadean" times, these scientists nonetheless obtained similar harvests of complex organic compounds. More recent considerations have the early Earth's atmosphere evolve from a composition dominated by H_2 and N_2 to one that again includes H_2O and CH_4 in response to an asteroid strike. Further chemical reactions lead to HCN, NH_3, and other biogenic compounds containing nitrogen (see Fig. 13.2).

W. H. Waller, *Crucibles of Creation*, Astronomers' Universe, https://doi.org/10.1007/978-3-032-17258-7_13

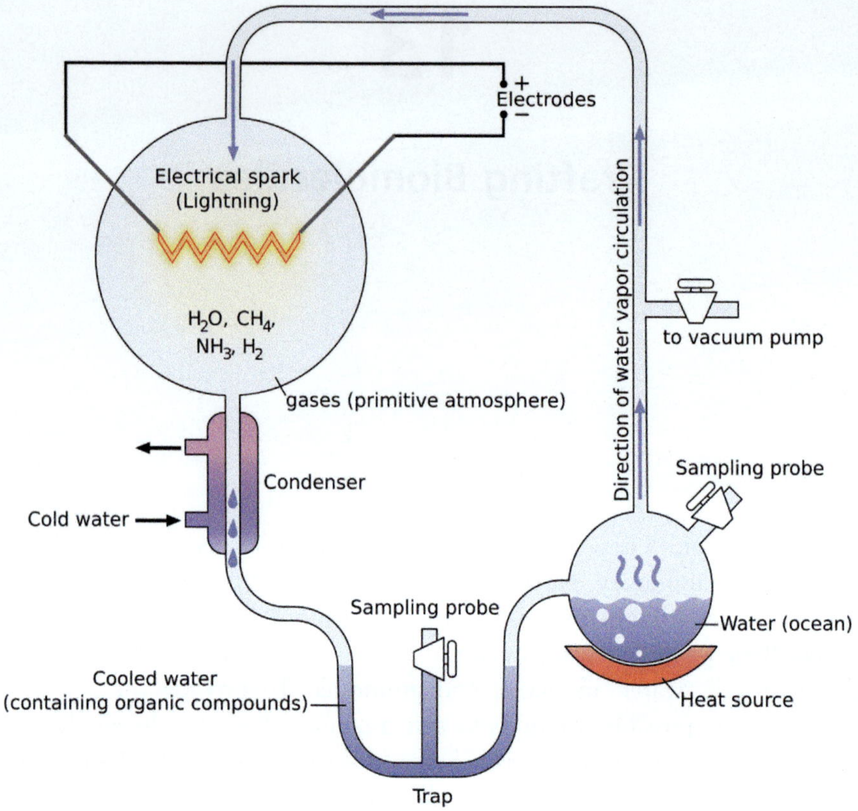

Electrodes
Electrical spark (Lightning)
H_2O, CH_4, NH_3, H_2
gases (primitive atmosphere)
Direction of water vapor circulation
to vacuum pump
Condenser
Cold water
Sampling probe
Water (ocean)
Heat source
Cooled water (containing organic compounds)
Trap

Fig. 13.1 Setup of the Miller-Urey experiment that yielded amino acids among other organic compounds (*credit* Wikipedia and https://creativecommons.org/licenses/by-sa/3.0)

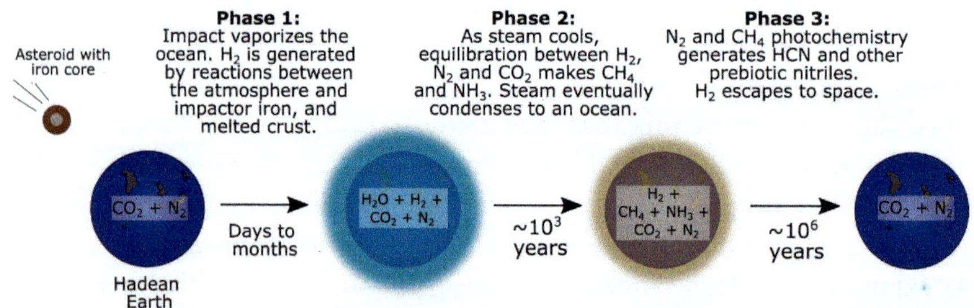

Fig. 13.2 Schematic evolution of early Earth's atmospheric composition resulting from one or more impacts from an iron-rich asteroid (*credit* Nicholas F. Wogan et al. (2023) Planet. Sci. J. vol. 4 p. 169, https://iopscience.iop.org/article/https://doi.org/10.3847/PSJ/aced83/meta, via Wikipedia and Creative Commons licence 4.0, https://creativecommons.org/licenses/by/4.0/deed.en)

Tracking Molecules on Multiple Scales

It is not within the scope of this book (nor in my realm of expertise) to fully elucidate the possible chemical pathways that could characterize the planetary atmospheres and oceans over cosmic time. Suffice to say that the rich brew of molecules that have been detected in galactic ecosystems certainly provides a surfeit of possible initial conditions. So, let's take stock of what we have found to date within GEs, proto-planetary disks, and exoplanetary atmospheres.

As the sampling of interstellar molecules in Table 7.1 attests, more than 330 different molecules have been detected within galactic ecosystems. Most of them were located within molecular clouds and circumstellar environments. Pretty much all of the elements involved with life on Earth (CHNOPS) are well-represented—but with a relatively poor showing of phosphorus that could imply a possible bottleneck. Molecules containing carbon, hydrogen, and oxygen abound, with a panoply of alcohols, sugars, ethers, aldehydes, and PAHs announcing their presence at radio and infrared wavelengths. Other identified molecules contain nitrogen (such as nitric oxide, nitrogen sulfide, ammonia, and hydrogen cyanide). These are especially vital to the business of building amino sugars, amino acids, and nucleic acids here on Earth.

Sensing Protoplanetary Disks

As of 2025, multi-wavelength observations of protoplanetary disks have yielded a harvest of about 40 different molecules. This count is down from the overall tally in GEs by more than a factor of ten. So, where have all the molecules gone? One possible explanation for the shortfall posits that many of the more volatile molecules have frozen onto dust grains in the disks and so have excluded themselves from emitting in the gas phase and thus being detected. Support for this idea comes from maps of protoplanetary disks in the light of different molecules, where strong radial variations are evident. The observed distributions of these various molecules likely depend on the molecules' respective responses to the central stars' energizing presence. Another possibility is that astronomers have just begun to resolve protoplanetary disks at infrared and millimeter wavelengths and so obtain sufficiently sensitive spectra (e.g. with *JWST* and *ALMA*). We literally remain in "a wait and see" situation.

Among the 40+ detections in protoplanetary disks, however, several versions of cyanides have piqued the interest of astrochemists (see Fig. 13.3).

These include various forms of hydrogen cyanide (HCN and DCN), cyanoacetylene (HC_3N), and methyl cyanide (CH_3CN). These cyanides along with ammonia could provide the essential feedstock for future nitrogenous macromolecules forming on the surfaces of emergent planets. How they would get transported to the protoplanets remains a highly speculative topic. Many of them would be destroyed at the equivalent of Earth's distance from the Sun, and so another formation site might be necessary. The outer Solar System, with its cache of icy comets, might provide just such a model birthplace. Once embedded within the cometary bodies, these molecules could then hitch rides to the inner parts of their respective exoplanetary systems and so get delivered to the surfaces of the young planets residing there. The long chain hydrocarbons found in molecular cloud cores likely won't survive the denser and more irradiated environments characteristic of protoplanetary disks. That would suggest having to re-start their synthesis within the disks and on any planets that may form therein.

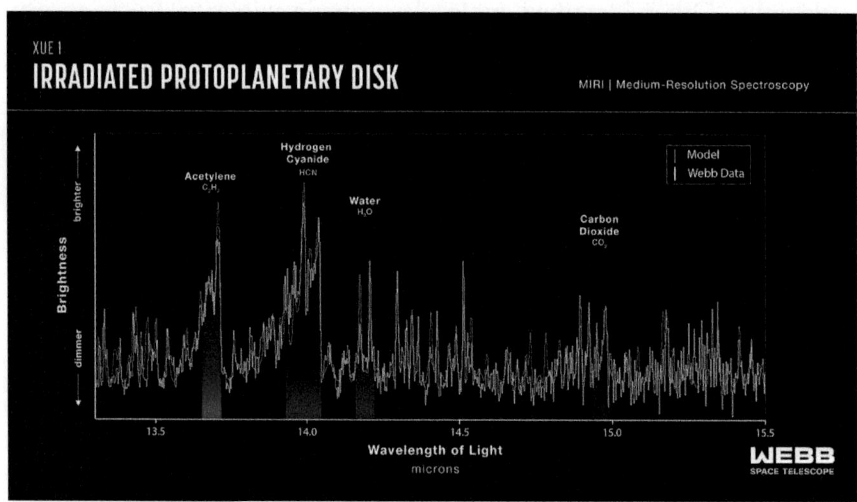

Fig. 13.3 Mid-infrared spectrum of the protoplanetary disk XUE 1, as obtained by *JWST*. Among the many emission features, those produced by carbon dioxide, water, hydrogen cyanide, and acetylene have been highlighted (*credits JWST*, NASA, ESA, CSA, STScI, J. Olmsted [STScI], M. C Ramírez-Tannus [Max Planck Institute for Astronomy])

Sensing Exoplanetary Atmospheres

Our ability to detect molecules in the atmospheres of exoplanets remains in a fledgling state. *HST* and *JWST* have successfully obtained revealing spectra of giant exoplanets transiting their host stars. Spectroscopic evidence for carbon monoxide (CO), carbon dioxide (CO_2), water (H_2O), sulfur dioxide (SO_2), hydrogen cyanide (HCN), methane (CH_4), ammonia (NH_3), and acetylene (C_2H_2) has been found in the atmospheres of these gas giants (see Fig. 13.4). The atmospheric situation with rocky planets, however, remains sullied by haze and clouds nulling the expected spectroscopic features. We can only guess that the atmospheric mixes might include what we see in our own inner Solar System—namely carbon dioxide, water, methane, diatomic hydrogen (in the distant past), diatomic nitrogen, diatomic oxygen, and ozone—the latter two molecules having biological origins. And then there are the possible techno-signatures that might one day be detected. These include waste heat and light produced by any exoplanetary civilizations, pulses of laser and radio emission at specific wavelengths, as well as "pollutants" such as nitrogen dioxide (NO_2) and chlorofluorocarbons (CFCs). Given all these prospects and challenges, it's an exciting time to be an astrochemist, with lots to anticipate in the coming decades.

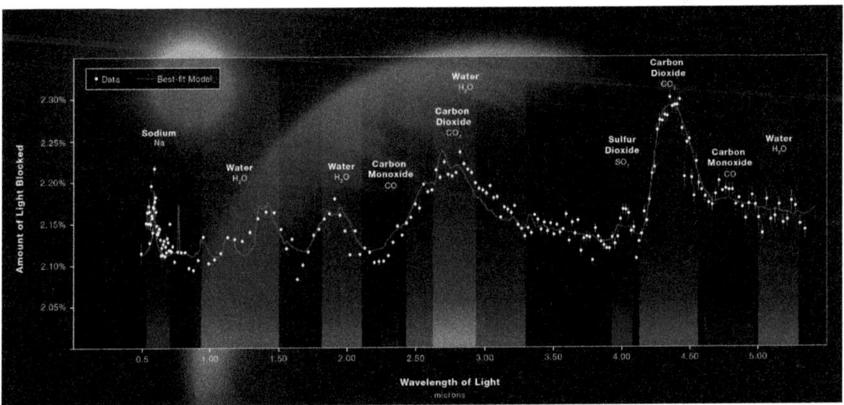

Fig. 13.4 Near-infrared spectrum of the giant exoplanet WASP-39b that resulted from the planet's transit in front of its star. The plot has been inverted, so that the atmospheric absorption features appear as "emission" features. The faint blue line denotes the output from an atmospheric model (*credits JWST,* NASA, ESA, CSA, J. Olmsted [STScI])

14

Coda

When we walk in the woods, stroll along the seashore, trek through desert badlands, or climb a mountain ridge, we can sense nature's agency and mystery. Somehow, we find our inner spirits resonating with the many marvels before us. So too, when we behold a starlit night sky, we can feel a strange affinity for all those inscrutable points of light (see Fig. 14.1). This sublime sense of awe should not be so surprising to us, as we are bearing witness to our natural birthright.

Over the past 10,000 years of human development, our inborn curiosity and technological advances have enabled us to first recognize and then deeply explore the nebular wellsprings of all this terrestrial and celestial foment. What we have found in these crucibles of creation further resonate with our notions of construction and destruction, dark secrets and fiery dramas—birth, transformation, death, and renewal. It's all there in these galactic ecosystems.

We have also learned that the story doesn't end with the stars, as most of them host planets of diverse variety and circumstances. If life is a planetary phenomenon, then certainly it has taken hold on at least a few of the planets that orbit the 100 billion stars in our Milky Way galaxy. We continue to find that our circumstances in the Solar System are not so unique. Other solar systems contain rocky planets in comfortable settings relative to their energizing stars. Water suffuses many of these exoplanetary realms, and complex chemistries abound.

In this book, I have tried to make the case that galactic ecosystems play an important role in fostering biogenic molecules. Moreover, their readily observable spectral line emission at multiple wavelengths makes them prime

W. H. Waller, *Crucibles of Creation*, Astronomers' Universe,
https://doi.org/10.1007/978-3-032-17258-7_14

Fig. 14.1 Water and wind-carved "hoodoos" in the badlands of New Mexico before a night sky replete with planetary, stellar, and nebular wonders (*credit* Courtesy of Marcin Zajac)

targets in the search for biomolecules. I would be remiss, however, to neglect the physical and chemical differences between galactic ecosystems writ large and the much smaller environments associated with protoplanetary disks or newborn planets. We are faced with a still unconstrained tale of chemical pathfinding—from the GEs through the protoplanetary disks to the surfaces of planets in habitable orbits around their host stars (see Fig. 14.2). What molecules get destroyed, which ones survive the voyage, and what can get made on-site? These vital questions of being and becoming continue to motivate the scientific nexus, where astrophysics, astrochemistry, and astrobiology converge. We have just begun to make the connections, so there's a whole lot more to explore. Surely, new insights will come as we continue to seek our cosmic roots within galactic ecosystems—the crucibles of creation.

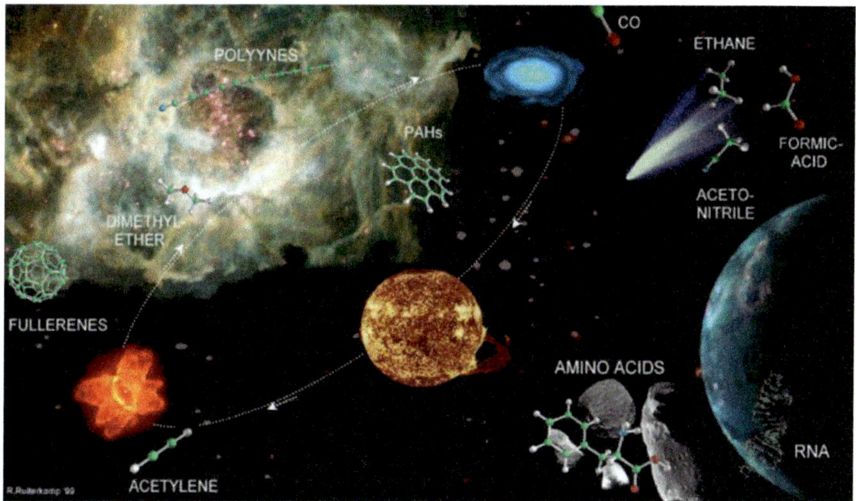

Fig. 14.2 Artist's conception of the chemical pathways in galactic ecosystems. To the upper left, a large cluster of stars has formed whose most massive members have ionized their nebular womb. To the upper middle, a young stellar object is energizing its disk of protoplanetary gas and dust. Following the arrows, the YSO evolves into a main-sequence star and then expires, creating a nebula full of new molecules and dust that mixes with future nebular sites of star and planet formation. To the right, a young planet receives biogenic molecules from infalling comets and meteorites. On the planet's surface, macromolecules such as RNA take form (*credits* R. Ruiterkamp—Ehrenfreund and Charnley (2000), in *Annual Review of Astronomy and Astrophysics*, with permission)

But after all, our dreams are all we are,
Fine dust from once exploding stars,
Imbuing breaths with mists set free,
From nebulae, to you and me.

– W. H. Waller (2025 CE)

Appendices

Appendix A: Nearby Molecular Clouds

In Chaps. 4, 5, and 10 we deeply dove into the Taurus, Corona Australis, and Rho Ophiuchi molecular clouds. Each of these clouds are thought to line the surface of the Local Bubble through which the Sun happens to be passing. In this Appendix, I am adding three more examples of molecular clouds for your perusal. The Pipe Nebula also appears to hew to the walls of this expanding bubble. It has been resolved into 130 molecular cloud cores, ripe with star-forming potential. The Perseus Molecular Cloud lies twice as far as these local clouds, perhaps as part of an adjacent bubble system. Then there is the Orion Molecular Cloud Complex—the nearest site of burgeoning starbirth activity. Here, the multiple stellar populations and gaseous remnants trace a 15 million-year-long legacy of starbirth and stardeath.

A-1: The Pipe Nebula

If you look towards the constellations of Sagittarius, Ophiuchus, and Scorpius on a clear dark summer night, you can discern regions of dark obscuration against the background of starlight. Amateur astronomers are especially adept at delineating these patches and tendrils with just a pair of binoculars. The Pipe Nebula (Barnard 59, Barnard 65-67, and Barnard 78) stands out, as its shape resembles that of a smoking pipe (see Fig. A.1).

Like the Taurus, Corona Australis, and Rho Ophiuchi molecular clouds, the Pipe Nebula resides on the surface of the same bubble that was blown by several supernova explosions millions of years ago (see Fig. 4.5). The

W. H. Waller, *Crucibles of Creation*, Astronomers' Universe,
https://doi.org/10.1007/978-3-032-17258-7

Fig. A.1 This view of the Milky Way looking toward the Galactic Center contains several outstanding objects, including the roseate **Lagoon Nebula** (M8) to the left, the **Rho Ophiuchi Molecular Cloud** to the right, and the **Pipe Nebula** near the center. At a distance of 542 light-years, the Pipe Nebula consists of myriad dark clouds and cloud cores in an extended system that spans about 46 light-years. (Credits: European Southern Observatory, S. Guisard)

expanding bubble's compressive dynamics likely plowed up the material that is now contained within these molecular clouds. By mapping the visual extinction produced by the obscuring dust in the Pipe Nebula, astronomers have identified about 130 dense cores. Moreover, the degree of extinction has enabled them to estimate the mass of molecular gas contained within these self-gravitating clumps. The number distribution of core masses (shown in Fig. A.2) is remarkably similar in shape to that of newborn stars, as surmised from observations of very young star clusters. The key difference in the core versus stellar mass distributions is an offset in masses by a factor of about 3–4. The astronomers have interpreted these similarities and factorial differences as arising from the same gravi-thermal physics that has resulted in a star-forming efficiency of about 30%.

A-2: The Perseus Molecular Cloud

I have chosen to highlight the Perseus Molecular Cloud, because of its prominence in the spatially-filtered map of far-infrared emission towards the outer Milky Way that was shown in Fig. 3.7. Located at a Galactic Longitude of about 160-deg, the cloud appears to have two components. The upper part at

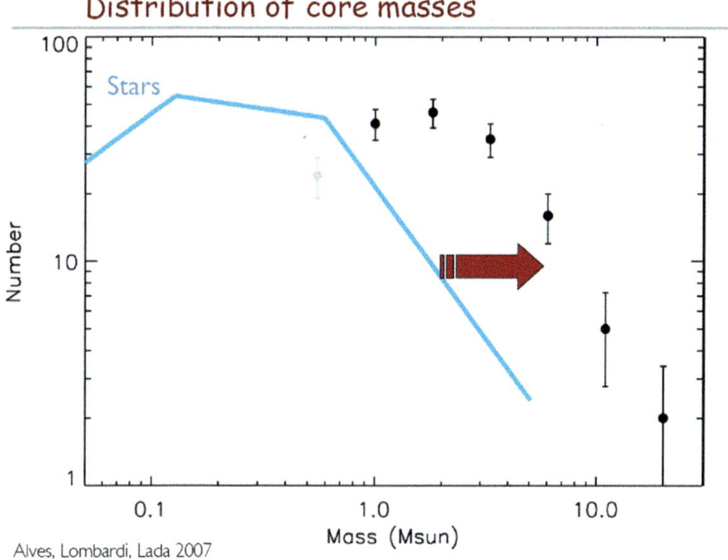

Fig. A.2 Number distributions of mass for molecular cloud cores in the Pipe Nebula versus newborn stars as observed in young star clusters. The similarity in shape suggests similar physics underlying the formation of both molecular cloud cores and the stars they produce (Credit: Courtesy of Charles Lada)

a Galactic Latitude of -12-deg corresponds to the California Nebula (NGC 1499)—an HII region that has been photoionized by the runaway O-type star Menkib (Xi Persei). The lower part at a Galactic Latitude of -20-deg comprises the main molecular cloud (see Fig. A.3). This cloud has an estimated distance of 1200 light-years, spans about 100 light-years, and contains roughly 10,000 solar masses worth of molecular gas.

The Perseus Molecular Cloud represents the currently active part of the Perseus OB2 stellar association. Containing 40 young massive stars along with thousands of other lower-mass stars, Per OB2 sprawls over 250 light-years of Galactic territory. Supernovae from this association are thought to have blown an even larger bubble that abuts the Local Bubble portrayed in Fig. 4.5. The young clusters shown in Fig. A.4 contain stars with ages that range over several million years, thus suggesting star formation having percolated over this period of time rather than having spawned in one big burst. Although these clusters do not contain any massive ionizing stars, they are chock-a-block with young stellar objects and associated outflow activity (see also Fig. 6.3). All this stellar and nebular foment serves as a prelude to the even more intense star-forming activity evident in the giant Orion Molecular Cloud Complex that will be reviewed next.

Fig. A.3 The **Perseus Molecular Cloud** as mapped at mid-infrared wavelengths by the *WISE* telescope in galactic coordinates. The field of view in this image is roughly 22-deg by 27-deg (460-ly by 560-ly at a distance of 1200 ly). The red-coded emission at 22-μm wavelength is from warm dust surrounding the hot O-type star, Menkib. The green-coded nebulosity at 12-μm wavelength above the star is from PAH molecules associated with the California Nebula (NGC 1499), a classic HII region. The greenish nebulosity towards the lower middle also traces PAH molecules. It includes the ring-like cloud Barnard 3 along with the star clusters IC 348 and NGC 1333 above and below B3. The Pleiades star cluster and coincident nebulosity appears to the lower left (Credits: Created using the World-Wide Telescope application and the *WISE* all-sky survey database, NASA/JPL-Caltech)

A-3: The Orion Molecular Cloud Complex

The easily recognized constellation of Orion the Hunter delineates the nearest complex of giant molecular clouds. At an estimated distance of 1200–1400 light-years, the clouds span more than 400 light-years and contain close to a million solar masses worth of cold molecular hydrogen gas. Parts of the cloud system are currently involved in vigorous star-forming activity, with all the nebular consequences that you might anticipate. Fig. A.5 provides a basic mapping of the molecular gas, courtesy of the mid-infrared emission from PAH molecules.

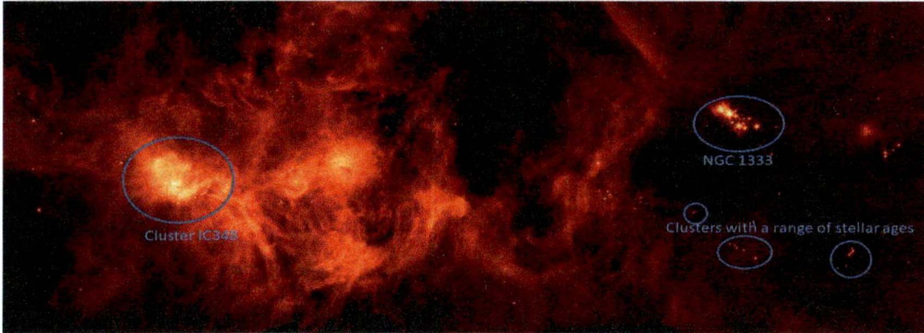

Fig. A.4 Mid-infrared closeup of the **Perseus Cloud** obtained by the *Spitzer Space Telescope* at a wavelength of 24 μm with newborn star clusters highlighted. This wavelength is most sensitive to emission from star-warmed dust. The ring-like nebulosity Barnard 3 shown in Fig. A.3 appears just to the right of the young star cluster IC 348. Other star clusters populate this 5.6-deg by 2.3-deg field of view, with NGC 1333 containing the greatest number of young stellar objects (Credits: NASA/JPL-Caltech, Luisa Rebull)

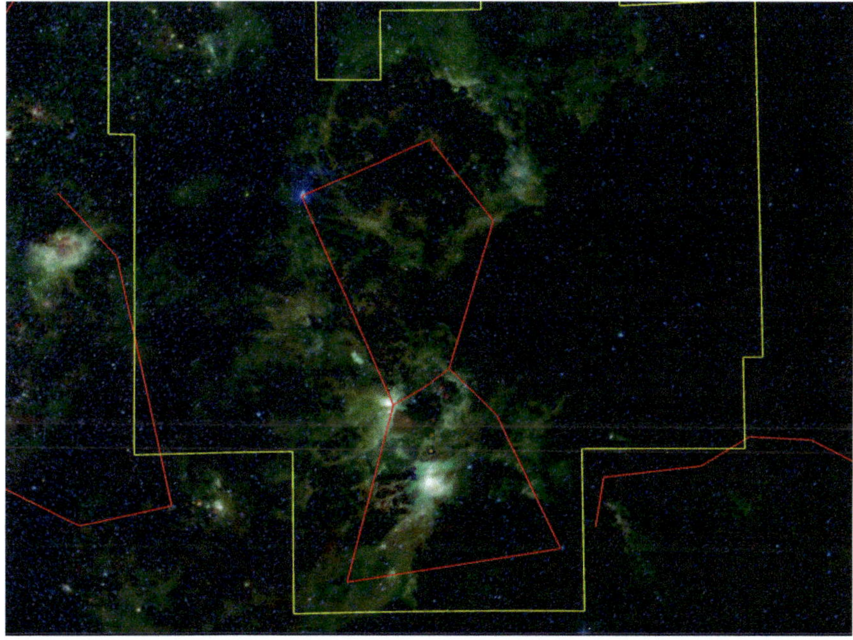

Fig. A.5 The **Orion Molecular Cloud**, as mapped by the *WISE* telescope at 4-μm (blue), 12-μm (green) and 22-μm (red). North is up and east to the left. The field of view on the sky is roughly 40-deg by 30-deg. The ubiquitous green-coded emission is tracing PAH molecules. The constellation's asterism is outlined in red, with its borders outlined in yellow. The Rosette Nebula in Monoceros appears to the far left (Credits: Created using the World-Wide Telescope application and the *WISE* all-sky survey database, NASA/JPL-Caltech)

Further insights can be drawn from Fig. A.6, where visible and far-infrared emitters are labeled and compared. While the visible view highlights the stars and ionized gas, the FIR view highlights the star-warmed dust and molecular gas. The nebulosity around Lambda Orionis (Meissa) is especially striking. Lambda Ori consists of two hot O8III and B0V stars. Their powerful ultraviolet emission has photoionized the surrounding gas which is seen in the red light of fluorescing hydrogen atoms. The UV emission and winds from these stars (perhaps assisted by a supernova explosion) have also plowed up a shell of much colder molecular gas. Barnard's Loop (Sh 2-276) is another fabulous consequence of the massive stars that have recently emerged from these clouds. It probably represents a shell that was inflated by one or more supernovae and is currently responding to the UV emission produced by the most recent crop of massive stars.

Throughout this giant galactic ecosystem, associations and clusters of newborn stars have contributed their energizing light, winds, and—ultimately—supernova blasts to the ambient nebulosity. There appears to be a spatial progression in the star-forming activity. The OB1a stellar association

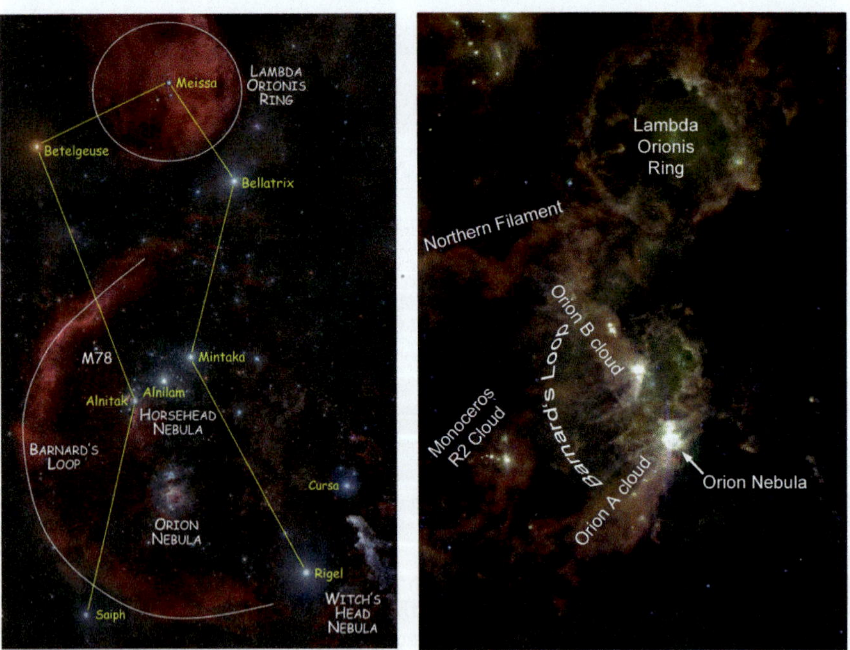

Fig. A.6 Comparison of the stellar and nebular emission in the direction of Orion at visible (left) and far-infrared (right) wavelengths (Credits: Adapted from [left] R. B. Andreo, deepskycolors.com and [right] NASA/IRAS/JPL-Caltech, T. Dame, via Wikipedia and Creative Commons Attribution-Share Alike 4.0 International license)

to the west (right-hand side of the above figures) has the greatest ages of 8–12 Myr. Then comes the OB1b stellar association centered on Orion's belt and the Horsehead Nebula with ages of 1.7–8.0 Myr. The OB1c association in Orion's sword has the youngest ages of 2–6 Myr. The cluster powering the Orion Nebula itself is even younger, with stars only 1 Myr old (see Fig. 8. 4). All told, about 5000 to 20,000 stars have formed in this region with 10–20 supernovae having popped off over the past 15 Myr (see J. Bally in B. Reipurth et al. [2008a]).

The Orion Molecular Cloud Complex has also served as a happy hunting ground for astrochemists eager to find and characterize molecules in diverse settings. They have not been disappointed, as the current tally of molecules features a rich suite of inorganic and organic species (see Table A.1). This listing neglects the even heavier PAHs that can contain up to 60 atoms.

Appendix B: Galactic HII Regions

There is no hard and fast rule for differentiating between classical HII regions, giant HII regions, and starbursts. As a rule of thumb, I would suggest that classical HII regions contain between one and ten ionizing O-type stars, giant HII regions contain 10–100 such stars, and starbursts host 100–1000+ of these stellar powerhouses. We have already explored the Orion Nebula (M42) which contains 3 ionizing stars (Theta 1 Orionis A [O6V] and the Theta 2 Orionis A system [O9.5VI, B0.7V, B5V]). Here, we will consider two other classic HII regions – namely the Trifid Nebula (M20) and its larger neighbor the Lagoon Nebula (M8). We also encountered in the main text the Eagle Nebula (M16), a giant HII region that contains about 14 O-type stars and 35 early B-type stars. Here, we up the stakes to revisit the Carina Nebula (NGC 3372)—a sprawling system that contains more than 65 O-type stars—as well as NGC 3603—a centrally-concentrated system with more than 50 O-type stars that provides a model for other even more powerful starbursting systems.

B-1: The Trifid and Lagoon Nebulae

There is nothing like scanning the constellation of Sagittarius with a decent pair of binoculars. On clear moonless summer nights away from city lights,

Table A.1 Listing of 43 molecules detected in the Orion Molecular Cloud Complex (D. Sahu et al. 2021 and references therein)

2 Atoms		3 Atoms	
Species	Regions	Species	Regions
SO	Orion-KL, Orion-A	OCS	Orion-KL, Orion-A
NO	Orion-KL	HNC	Orion-KL, Orion-A
CS	Orion-A	SO_2	Orion-A
CO	Orion-A	HDO	Orion-A
SiS	Orion-A	HCO^+	Orion-A
PO	OMC-1	HCS^+	Orion-A
PN	OMC-1	HCO	Orion-A
		HCC??	Orion-A
		HCN	Orion-A
		HNC	Orion-A
		C_2S	Orion-KL
		SiC_2	Orion-KL
4 Atoms		**5 Atoms**	
Species	**Regions**	**Species**	**Regions**
H_2CO	Orion-KL	HC_3N	Orion-KL, Orion-A
HNCO	Orion-KL, Orion-A	H_2CCO	Orion-KL, Orion-A
H_2CS	Orion-KL	$c\text{-}C_3H_2$	Orion-KL
HNCS	Orion-KL	HCOOH	Orion-KL
		NH_2CN	Orion-KL
6 Atoms		**7 Atoms**	
Species	**Regions**	**Species**	**Regions**
CH_3OH	Orion-KL, Orion-A	CH_3CHO	Orion-KL
CH_3CN	Orion-KL, Orion-A	CH_3CCH	Orion-KL, Orion-A
CH_3SH	Orion-KL	CH_2CHCN	Orion-KL, Orion-A
NH_2CHO	Orion-KL	HC_5N	Orion-KL
8 Atoms		**9 Atoms**	
Species	**Regions**	**Species**	**Regions**
$HCOOCH_3$	Orion-KL, Orion-1	CH_3OCH_3	Orion-KL, Orion-A
		CH_3CH_2CN	Orion-KL, Orion-A
		HC_7N	Orion-KL
		CH_3CH_2OH	Orion-KL

OMC-1 refers to the molecular cloud directly behind the Orion Nebula. Orion-KL refers to the Kleinmann-Low Nebula, a young star-forming region about 5-arcmin to the northwest of the Trapezium star cluster in the Orion Nebula (and perhaps part of OMC-1) that is currently erupting with multiple energetic outflows

you will find all sorts of "faint fuzzies" amid the dense star fields.[1] First on the list is the Lagoon Nebula (M8), near the spout of the teapot asterism that makes up part of Sagittarius (see Fig. 3.5). Nearby, is the smaller Trifid

[1] Besides the HII regions mentioned here, other "faint fuzzies" in Sagittarius include a rich assortment of open and globular star clusters—the vestiges of prior starbirth activity in long-lost galactic ecosystems.

Fig. B.1 Visible view of the field in Sagittarius, where the **Lagoon Nebula** (M8) appears toward the bottom of the field, the **Trifid Nebula** (M20) near the top, and **Simeis 188** (Sharpless 29) towards the left. This color image was obtained through blue, green, red, and H-alpha filters and then combined and presented with a "natural" palette. The angular field of view is 3.5-deg by 2.7-deg which, at a distance of about 4500-ly translates to a physical extent of 275-ly by 212-ly. Throughout the field, dark tendrils of obscuring dust delineate the remnant molecular clouds from which the newborn hot stars and ionized nebulae emerged. The dominant red glow is from the ionized hydrogen in these HII regions. The blue nebulosity near the top of the Trifid Nebula is from non-ionizing blue starlight scattered off grains of dust in the nebula (Credit: Courtesy of Phil Orbanes, Cosmic Mariner Observatory, Gloucester Area Astronomy Club, https://gaac.us/gallery-2/phil/phil2#!prettyPhoto[https://phil2]/81/)

Nebula (M20)—and further along the same swath of the Milky Way—the Omega/Swan Nebula (M17) and Eagle Nebula (M16). Fig. B.1 provides a closer view of the Trifid and Lagoon Nebulae in the context of other bright and dark nebulae that adorn this part of the Sagittarius spiral arm, some 4500 light-years away.

Because of its relative isolation, small size, and circular symmetry, the **Trifid Nebula** provides a valuable laboratory for investigating stellar-nebular feedback in galactic ecosystems. The O7V star HD 164492 is largely responsible for photo-ionizing the ambient gas. The resulting HII region's size of about 18 light-years is consistent with expectations based on the single star's EUV luminosity (see Appendix E-5). The ionizing star is joined by 4 other non-ionizing stars along with 4 infrared sources as part of a small central cluster. Elsewhere within the nebula, the infrared *Spitzer Space Telescope* identified 150 YSOs spanning the Class 0, I, and II evolutionary stages (see Ch. 6). These stellar newborns indicate a very young star-forming region with a characteristic age of less than a million years. At millimeter wavelengths, the *IRAM* telescope has found 33 dusty condensations, about half of which are currently forming stars. Elsewhere, the *HST* has resolved multiple cometary globules. Shaped by photo-evaporative processes, these evaporating gaseous globules ("eggs") contain protoplanetary disks—similar to those previously observed in the Orion Nebula and Eagle Nebula (see Ch. 6 and Ch. 8, respectively).

Dust lanes trisect the Trifid Nebula – giving it a unique appearance and its evocative name. When observed at mid-infrared wavelengths, these dust lanes glow in the light of star-warmed dust (see Fig. B.2). We can now see that the interior of the nebula hosts narrow filaments of gas and dust, whose compression bespeaks shocking conditions. Throughout the nebula, emission from excited PAH molecules delineate the working surfaces of the remnant molecular cloud.

The **Lagoon Nebula** (M8 or NGC 6523) is considerably larger, as it is responding to a more robust population of newborn stars (see Fig. B.3). Towards the east (left end) of the nebula, the open cluster known as NGC 6530 contains thousands of young stars, 3–5 of which have O-type classifications. This cluster is thought to be 1-3 Myr old. At least part of the cluster is expanding. Towards the west (right) part of the nebula, are two notable stars. Herschel 36 appears brightest, with a O7V spectral classification. It is surrounded by the dusty Hourglass Nebula whose twin openings lend the nebula its name. 9 Sagittarii can be seen to the east (left) of Herschel 36. It consists of a binary system with O6.5V and O7.5V classifications.

The Lagoon's appearance at mid-infrared wavelengths startles with its disorienting complex of nebulosity emitting in the light of star-warmed dust and photo-excited PAH molecules (see Fig. B.4). The working surfaces are not as well defined as in the Trifid Nebula, perhaps due to the scattered distribution of exciting stars. However, further scrutiny of the visible and mid-IR images reveals spatial correspondences that confirm the salience of the PAH emission as delineating physical vanguards between the molecular and ionized phases, ie. PDRs. Meanwhile, the presence of numerous dark globules and

Fig. B.2 Visible and mid-infrared views of the **Trifid Nebula**. North is up and east to the left. At an estimated distance of about 4500 ly, the 17-arcmin by 28-arcmin fields of view each translate to a physical extent of 22-ly by 36-ly. *Left*—The view at visible wavelengths highlights the central ionizing star, fluorescing hydrogen gas, and triplet of obscuring dust lanes (Credit: National Optical Astronomy Observatory [NOAO], now called NOIRlab) *Right*—The similarly scaled view at mid-infrared wavelengths reveals emission from star-warmed dust at 24-μm wavelength (depicted in red), energized PAH molecules at 8-μm wavelength (in green), and stars at 4.5-μm wavelength (in blue). Near the top of the field, the reflection nebulosity illuminated by the A-type star HD 164514 shows emission from both PAHs and dust (Credits: *Spitzer Space Telescope,* NASA/JPL-Caltech/J. Rho [SSC/Caltech])

sub-mm emitting molecular clumps underscore the ongoing starbirth activity within this system (cf. Tothill et al. in B. Reipurth 2008b). These multi-wavelength revelations bode well for interpreting similar observations of even larger galactic ecosystems, as will be explored next.

B-2: The Carina Nebula

There is so much going on in the Carina Nebula (NGC 3372) that it is hard to know where to begin. Examination of the wide-field views in Fig. 3.8

Fig. B.3 Closeup of the **Lagoon Nebula** at visible wavelengths, showing a loose cluster of hot young stars amid a complex skein of fluorescing hydrogen gas and obscuring dust. North is up and east to the left. The angular field of view is 58-arcmin by 40-arcmin which, at a distance of 4500-ly, translates to a physical scene spanning 76-ly by 52-ly. The image was obtained through multiple filters with the *VLT Survey Telescope* atop ESO's Paranal Observatory in Chile. It is presented with a "natural" color palette (Credit: European Southern Observatory—VPHAS + team)

shows that it is one of the brightest HII regions in the sky. It is also associated with multiple far-infrared sources along with possible outflows of diffuse FIR-emitting dust. Further study of this giant HII region confirms its dominant status within the Milky Way galaxy. Its angular extent of 120-arcmin by 120-arcmin outspans the Orion Nebula in the sky by a factor of four. Moreover, it outshines the Orion Nebula by a factor of 16. At an estimated distance of 9,000 ly, its angular size translates to a physical size of 300 ly—more than ten times larger than the Orion Nebula. To view the Carina Nebula naked eye from a dark and dry site in the southern hemisphere is a special treat that I highly recommend!

Telescopic images of the nebula reveal a complex scene, full of dark lanes of obscuring dust, bright rims, and rich star fields (see Fig. B.5). The most energetic regions include the eruptive binary star system Eta Carina, the Keyhole Nebula beside Eta Car that resulted from the system's last major eruption around 1840 CE, the Trumpler 14 star cluster about 10 arcmin to the northwest of Eta Car, and the Trumpler 16 star cluster whose members are

Fig. B.4 Mid-infrared view of the **Lagoon Nebula** centered on Herschel 36 and the Hourglass Nebula. North is up and east to the left. The angular field of view is 36-arcmin by 29-arcmin, about 60% that in Fig. B.3. Emission from dust at 24-micron wavelength is portrayed in red, PAHs at 5.8- and 8.0-μm wavelengths in green, and stars at 3.6- and 4.5-micron wavelengths in blue (Credit: *Spitzer Space Telescope*, NASA/JPL-Caltech)

scattered throughout much of the nebula's southern half. All told, about 65 O-type stars currently populate the Carina Nebula. Their combined ionizing radiation outshines that in the Orion Nebula by a factor of 140.

An entirely different visage presents itself at mid-infrared wavelengths (see Fig. B.6). Instead of ionized hydrogen gas dominating the scene, light from UV-irradiated PAH molecules reveal the nebula's working surfaces – where the molecules are getting photo-dissociated and then photoionized by the EUV onslaught. The overall appearance is analogous to a geode that has been cut open. Nebular pillars point towards Eta Car, with evaporating gaseous globules ("eggs") and incubating stars evident at their tips.

But wait, there's more! This giant HII region manifests an explosive personality at X-ray wavelengths (see Fig. B.7). A two-week-long imaging exposure with the *Chandra X-ray Telescope* has revealed diffuse emission at multiple X-ray energies. The X-ray glow likely comes from million-degree

Fig. B.5 Visible-light view of the **Carina Nebula**. The angular field spans 1-deg, with north up and east to the west. The brightest source above the field's center is the eruptive binary star system Eta Carinae. The Keyhole Nebula appears to its right. Dark globules are evident along the dust lane to the upper right and the roseate nebulosity to the lower left. Compare this view with the closeup in Fig. 3.11 (Credit: European Southern Observatory—*VLT Survey Telescope*)

gas that was shock heated by recent supernova blasts. Our experiences with isolated supernova remnants show that the torrid gas that produces this sort of high-energy emission dissipates rather quickly—well short of a million years. That means we are witnessing repercussions from the latest salvo of stellar explosions to wrack the Carina Nebula. Besides detecting the hot gas, *Chandra* recorded X-ray emission from the churning atmospheres of more than 14,000 massive young stars. So, more fireworks may lie ahead.

Fig. B.6 Mid-infrared view of the **Carina Nebula** as obtained with the *Spitzer Space Telescope*. The angular field spans 2.1-deg, with north directed 28-deg from the vertical. Here, the color coding is different from that used in *Spitzer* images elsewhere in this book. PAH emission at 8-micron wavelength is represented in reddish tones, while ionized hydrogen emission within the 4.5-μm bandpass is represented in green tones. (Credit: *Spitzer Space Telescope*, NASA/JPL-Caltech)

Our homage to the Carina Nebula ends with two of its iconic residences—both imaged in exquisite detail by the *Hubble Space Telescope*. We begin with the massive binary star system Eta Carina. When Eta Car last erupted in 1837–1843, it brightened by a factor of 100 to become the second most luminous "star" in the sky (after Sirius). Today, it shines at a visual magnitude of 7.6—just below naked-eye visibility. Considering its distance of 9000 ly, however, its intrinsic luminosity is more like 4 million times the radiative output of the Sun. All this power is coming from two massive stars in highly

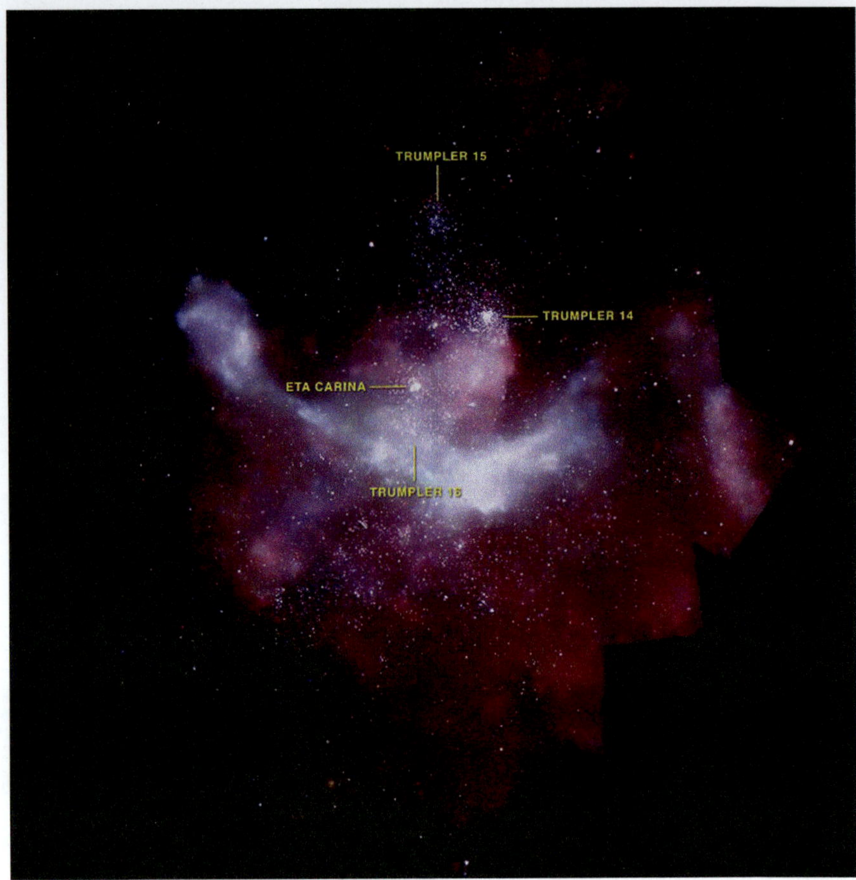

Fig. B.7 The **Carina Nebula** in X-rays. The angular field of view spans 1.7-deg, with north is up and east to the left. Diffuse emission is evident at low X-ray energies (shown in red) and high X-ray energies (shown in blue). X-ray emission from the active atmospheres of myriad young stars pepper the scene (Credits: *Chandra X-ray Telescope,* NASA/CXC/PSU/L.Townsley et al.)

elongated orbits about their common center of mass—a luminous blue variable (LBV) star that is entwined with an O-type star. Every 5.5 years, these two monsters come perilously close to one another, causing their mighty winds to collide. That's when the eruptive behavior tends to occur. What we see today is the residue of the last major eruption around 1840. The so-called Homunculus Nebula is characterized by a bipolar pair of bubbles that form the "mini-human" (or homunculus) with some strange features near the constricted "waist" (see Fig. B.8—left). More strong eruptions are predicted in coming decades to centuries. Then there is the likely prospect that the

Fig. B.8 *Left—HST* image of the nebular remnant surrounding the **Eta Carina** binary star system that erupted around 1840. (Credits: *HST,* NASA, ESA, Judy Schmidt—Creative Commons 2.0). *Right*—The **Flippin Bird (aka Defiant Finger) Nebula** (Credits: *Hubble Space Telescope,* NASA, ESA, N. Smith [University of California, Berkeley], and the Hubble Heritage Team [STScI/AURA])

most unstable of the two stars (the LBV) will go supernova sometime within the next few thousand years. Best stay tuned!

Amid the many globules of dense gas and obscuring dust in the Carina Nebula, the "Flippin Bird" Nebula exemplifies the genre of "Bok globules" that are being illuminated and photo-eroded by nearby hot stars (see Fig. B.8—right). This globule, in particular, evokes a certain irreverent sense of humor that seems to permeate the heavens. Many astronomers have felt themselves the butt of similar celestial jests from time to time.

B-3: NGC 3603

As shown in Figs. 3.8 and B.9, the Carina Nebula shares its stretch of the southern Milky Way with several other giant HII regions. These include RCW 49, NGC 3293, NGC 3324, NGC 3572, NGC 3576 and NGC 3603 – the latter being the most concentrated system of massive stars and responding nebulosity within the Galaxy. NGC 3603 is more than twice as far as the Carina Nebula, and so its relatively modest appearance among its neighboring HII regions belies a true giant among giants (see Fig. B.10).

A closer view of the mid-IR emitting nebulosity shows scads of emission from star-warmed dust and excited PAH molecules (see Fig. B.10). The brightest emitting region traces the realm dominated by the starburst cluster NGC 3603 (see Fig. B.11). Containing 6 hot and windy Wolf-Rayet stars

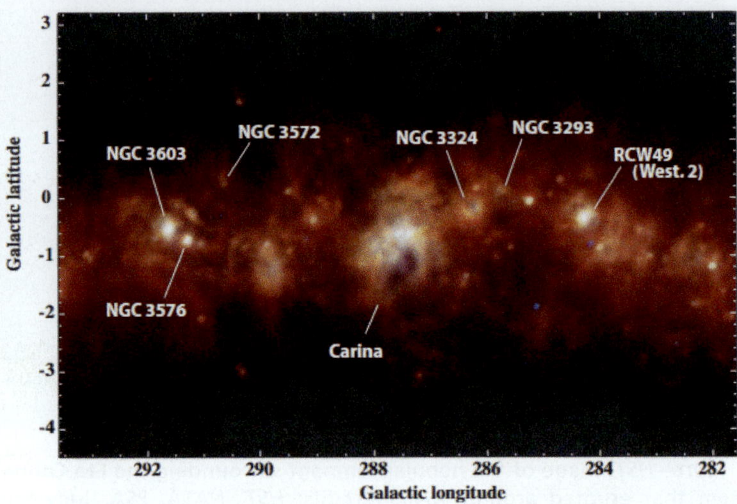

Fig. B.9 Far-infrared view of the Milky Way in the neighborhood of the southern constellation Carina. This mapping was obtained by the *Infrared Astronomy Satellite (IRAS)* at 12, 25, 60, and 100-micron wavelengths, with the reddest colors depicting emission at the longest wavelengths. The angular field of view is approximately 11-deg by 7-deg. Emitting regions are at differing distances, with the Carina Nebula 9000 ly away and NGC 3603 much farther at a distance of 20,000 ly (Credits: *IRAS*, NASA/JPL-Caltech, from N. Smith and K.J. Brooks in B. Reipurth 2008b)

and more than 50 blazing O-type stars, this incredible stellar cache has had its way with its birth cloud. In resistive response, two enormous pillars jut towards the cluster, their exposed surfaces succumbing to photoevaporation by the hot stars. Just beyond the cluster, the blue supergiant star Sher 25 is literally blowing its stacks in an eruption that could presage an imminent supernova. Other supernovae likely have exploded within this hotbed, as it glows brightly in X-rays. With NGC 3603, we can begin to understand what might be driving the even more powerful starburst activity in galaxies well-beyond our Milky Way.

Fig. B.10 Mid-infrared view of the nebulosity associated with **NGC 3603** as obtained with the *Wide-field Infrared Survey Explorer (WISE).* The angular field of view is approximately 2-deg by 2-deg which at a distance of 20,000 ly corresponds to a physical extent of 700 ly. Stellar emission at 3.4 and 4.6-μm wavelength is depicted in blue, emission from excited PAHs at 12-μm wavelength is shown in green, and emission from star-warmed dust at 22-μm wavelength is presented in yellow–red colors. The brightest central region is shown at visible wavelengths in Fig. B.11 (Credits: NASA/JPL-Caltech)

Fig. B.11 The starburst cluster **NGC 3603**, as imaged by the *Hubble Space Telescope* in 1999 with the Wide-field and Planetary Camera 2, the first camera to incorporate corrective optics for the spherical aberrations in the telescope's mirror. The angular field of view is 2.7 arcmin by 2.7 arcmin which translates to a physical field spanning 16 ly. North is up and east to the left. The visible-light image is presented with an approximately "natural" color palette (Credits: *HST*, NASA-ESA, Wolfgang Brandner [JPL/IPAC], Eva K. Grebel [Univ. Washington], You-Hua Chu [Univ. Illinois Urbana-Champaign])

Appendix C: Starbirth and Starburst Activity in Other Nearby Galaxies

Our Milky Way galaxy—along with the Andromeda Galaxy (M31)—comprise the most massive members of our Local Group—itself consisting of about 40 other much smaller galaxies. Several of these galaxies sport active galactic ecosystems worthy of our attention. Among them, the galaxies currently hosting powerful starburst activity include the Large Magellanic Cloud (LMC) and the Triangulum Galaxy (M33). We end this section with a brief look at active spiral galaxies beyond the LG.

C-1: The Large Magellanic Cloud (LMC)

The Large and Small Magellanic Clouds are scooting past our Milky Way like two playful schoolchildren who are just now responding to their teacher's calls. Over the past 400 million years, they have journeyed especially close to the Milky Way, whose tidal influence has incited vigorous starbirth and starburst activity within these dwarf systems. The Large Magellanic Cloud (LMC), in particular, currently hosts hundreds of HII regions over its 30,000-ly expanse (see Fig. C.1). These range in size from a few tens of light-years for most of them to more than a thousand light-years for the Tarantula Nebula—the largest HII region in the Local Group.

The nebular wellsprings of all this stellar birthing become readily apparent at mid-infrared wavelengths, where energized molecular clouds radiate in the light of PAH molecules, and where the star-warmed dust glows at temperatures of 10–100 K (see Fig. C.2). Here, the Tarantula Nebula becomes part of a much larger swath of interstellar gas and dust. Near the top of the field, a ring-like assortment of nebulae trace the supershell LMC 4 that is associated with a diadem of HII regions, as seen in Fig. C.1. This shell lies amid a froth other nebulae, resembling what we found in the Milky Way (see Figs. 3.5–3.8).

Before we pay our respects to the Tarantula Nebula, let's first consider a far more modest emission nebula known as **N70** (see Fig. C.3) Its nearly perfect spherical symmetry has attracted study by many astronomers including yours truly. What produced such a sublime nebula? Some astronomers have proposed that it represents a superbubble blown by the massive stars contained within. Others look to the extreme UV light from these stars impacting its surroundings, while still others claim supernova origins. The massive stars are certainly there, with at least 10 B-type stars and 6 O-type stars currently inhabiting the 300-ly expanse. Their presence indicates

Fig. C.1 The **Large Magellanic Cloud** at visible wavelengths with its many HII regions glowing in the red light of ionized hydrogen. This view spans about 10°, which at a distance of 160,000 ly translates to a physical extent of roughly 30,000 ly. The Tarantula Nebula (aka 30 Doradus, NGC 2070) appears to the left (east) of this image. The galaxy's stellar bar runs from lower left to upper right (Credits: European Southern Observatory, Courtesy of Robert Gendler and Josch Hambsch)

a minimum age of 5 Myr for the system. By imaging N70 in the light of ionized hydrogen (HII), sulfur ([SII]) and oxygen ([OIII]), my colleagues and I found remarkable enhancements in the [SII] emission within the shell and in the [OIII] emission near the center of the nebula (see Fig. C.4). These enhancements indicate collisional shocks exciting the gas within the shell, with scorching EUV radiation ionizing the gas near the hottest stars in the interior.

We interpreted these findings as indicating a mix of photoionization and shocks energizing the gas in N70. The shocks, in turn, could arise from a mix of strong stellar winds and prior supernova blasts. So, even in this relatively simple nebula, we can see the full gamut of stellar energetics impacting their surroundings. I remain humbled by the results of this one deep dive into the many wonders contained within the Large Magellanic Cloud.

Fig. C.2 The **Large Magellanic Cloud** at mid-infrared wavelengths shows the nebular context of the massive star-forming regions seen in Fig. C.1. Here, the field of view is a somewhat truncated 7.5-deg by 7.5-deg. The PAH line emission at 8-micron wavelength is coded green, while the dust continuum emission at 24-micron wavelength is coded red. (Credits: *Spitzer Space Telescope*, NASA/JPL-Caltech)

OK, onward to the star of the show – the **Tarantula Nebula**! This extraordinary HII region dwarfs any other HII region in the Local Group of galaxies (see Fig. C.5). The hot O-type stars contained within it dish out a thousand times more ionizing UV radiation than those within the Orion Nebula. The nebula's resulting size tops out at close to 1500 light-years. Were it situated at the distance of the Orion Nebula, it would cover a third of the sky! Truly, we are dealing with a major starburst here. Over its 25 Myr history of stellar birthing and dying, it has left behind hundreds of hot massive stars and more than 3600 point sources radiating in X-rays. These represent a mixed bag, including the hot massive stars, still-gestating stars, and

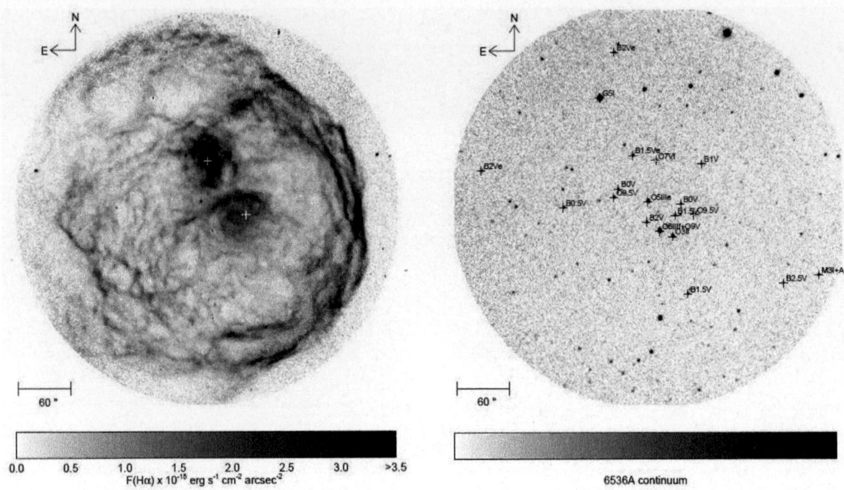

Fig. C.3 Emission-line and red-continuum images of the ionized bubble **N70** in the LMC. *Left*—The H-alpha emission from ionized hydrogen The nebulosity shows complex corrugations consistent with powerful dynamics at work. *Right*—The red continuum emission highlights the stellar population that includes several B-type and O-type stars. Field of view is about 7 arcminutes or 325-ly at the distance of the LMC (Credits: adapted from B. P. Skelton et al., PASP, 1999, Vol. 11, p. 465)

interacting binary star systems. Pervading the starburst, diffuse X-ray emission announces the presence of million-degree gas. Its appearance from our perspective suggests a maple leaf or bouquet, whose components could have arisen from the merging of supernova-driven superbubbles.

At optical and infrared wavelengths, the emission traces the cooler and denser "skeleton" of the nebula. And at radio wavelengths, we can see tendrils of cold star-forming gas remarkably close to R136—the densest and most powerful star cluster in the Local Group.

Looking closer, the R136 cluster breaks up into a bevy of hot massive stars (see Fig. C.6). Some of these barnburners embody more than 150 Suns worth of mass—pushing the theoretical limit set by their extraordinary luminosities (see Appendix E-7). Prior observations had indicated the presence of even more massive stars, with each star containing several hundred Suns worth of blazing gas. However, more recent observations at ever greater angular resolution have shown these monster "stars" to consist of multiple stars, each with masses no greater than 200 solar equivalents. In R136, we now appear to have hit the true limit of stellar mass.

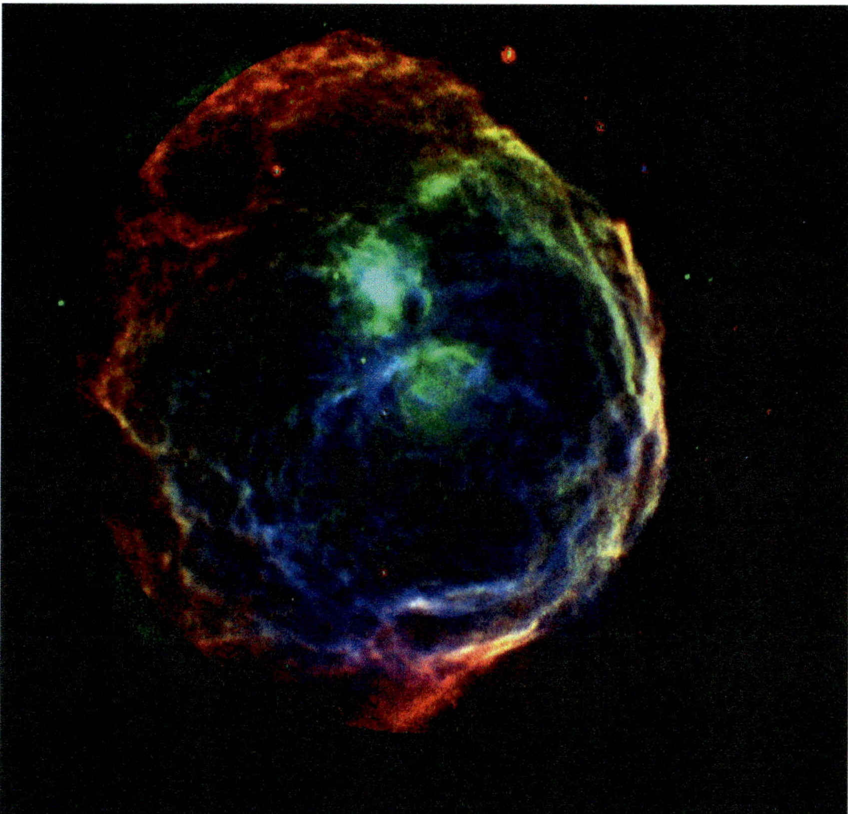

Fig. C.4 Emission-line image of the ionized bubble **N70**. The color coding presents the [SII] emission in red, H-alpha emission in green, and [OIII] emission in blue. Strong enhancements in the [SII] emission are evident in the outer shell, with [OIII] emission enhanced in the interior close to the ionizing stars (Credits: adapted from B. P. Skelton et al., PASP, 1999, Vol. 11, p. 465)

C-2: The Triangulum Galaxy (M33)

On clear and dark autumn nights, the Triangulum Galaxy (aka M33, NGC 598) appears as a faint haze, about 10 degrees to the southeast of the larger Andromeda Galaxy (M31). Binoculars will confirm its presence but that's about it. If you have access to a decent amateur telescope, you can see some details of this spiral galaxy, including a few of its giant HII regions. Long telescopic exposures with digital cameras, however, will show this galaxy to be churning with myriad star-forming regions (see Fig. C.7).

As early as the 1970s, astronomers had identified upwards of 1,500 HII regions in deep photographic plates and, later, digital images of this galaxy. Intermediate in size between the Milky Way Galaxy and the Large Magellanic

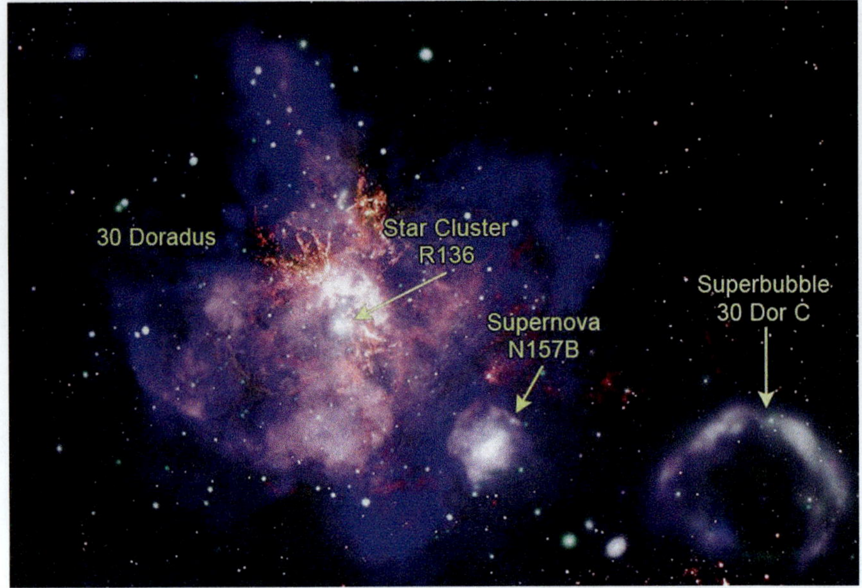

Fig. C.5 The **Tarantula Nebula** (aka 30 Doradus, NGC 2070) at multiple wavelengths. The field of view spans 30 arcmin, or 1400 light-years at the distance of the LMC. The blue and magenta colored regions trace X-ray emission from million-degree gas. Stellar sources of X-rays are shown in green. The reddish-coded regions trace both optical emission from ionized gas and infrared emission from dust. The orange-coded protrusions indicate the presence of radio-emitting tendrils of dense star-forming gas. Supernova N157B and Superbubble 30 Dor C shine brightly in X-rays, adding to the busy scene (Credits: X-ray: NASA/CXC/Penn State Univ./L. Townsley et al.; Infrared: NASA/JPL-CalTech/SST; Optical: NASA/STScI/HST; Radio: ESO/NAOJ/NRAO/ALMA; Image Processing: NASA/CXC/SAO/J. Schmidt, N. Wolk, K. Arcand)

Cloud, M33 presents a fascinating laboratory for studying recent starbirth activity within a spiral galaxy. Its brightest HII region, NGC 604, is the second most powerful star-forming region in the Local Group with only the Tarantula Nebula outshining it. But there are plenty of other giant HII regions and hosting galactic ecosystems to ponder. We can begin to do this by studying the far-infrared emission from the associated dust, as shown in Fig. C.8.

Here, we can see that the emitting dust nicely delineates the galaxy's raggedy spiral arms. The dust provides a handy proxy for the clouds of molecular gas—ground zero for the next generation of stars. The spiral arms are not fixed, but rather represent wave-like traffic jams in the rotating and shearing disk. Similar dynamics are probably playing out in our Milky Way Galaxy.

Further insights can be gleaned by comparing the light from multiple phases of interstellar matter in M33. In Fig. C.9, emission from ionized

Fig. C.6 At the heart of the Tarantula Nebula, the **R136 cluster** of hot massive stars plays a dominant role in energizing much of the nebula. This visible-light view by the *HST* resolves the cluster into its component stars, many of which hold the record for being the most massive known. The field of view is about 3 arcminutes which translates to a physical expanse of about 140 light-years at the distance of the LMC (Credits: *HST,* NASA, ESA, F. Paresce [INAF-IASF, Bologna, Italy], R. O'Connell [University of Virginia, Charlottesville], and the Wide Field Camera 3 Science Oversight Committee)

hydrogen, molecular PAHs, and granular dust are portrayed in corresponding colors of red, blue, and green. The emitting hydrogen gas clearly extends farther from the galaxy's center than the emitting PAHs and dust. This effect is likely due to the lower abundance of elements heavier than helium that has been observed in the outskirts of M33. The lower abundances, in turn, have squelched the formation of carbon-based PAHs and silicon-based dust grains out there. We can now appreciate the importance of elemental abundance

Fig. C.7 Visible view of the **Triangulum Galaxy** (M33) showing a disk of stars that is punctuated with many red H-alpha emitting HII regions along its ragged spiral arms. The image's angular field of view is 57 arcminutes by 68 arcminutes, which corresponds to dimensions of 46,000 ly by 55,000 ly at the galaxy's estimated distance of 2.8 million ly. To the upper left (northeast), the supergiant HII region NGC 604 outshines all of the other HII regions (Credits: ESO—*VLT Survey Telescope*)

in fostering galactic ecosystems that have the "right stuff" for developing complex molecules essential to the chemistry of life.

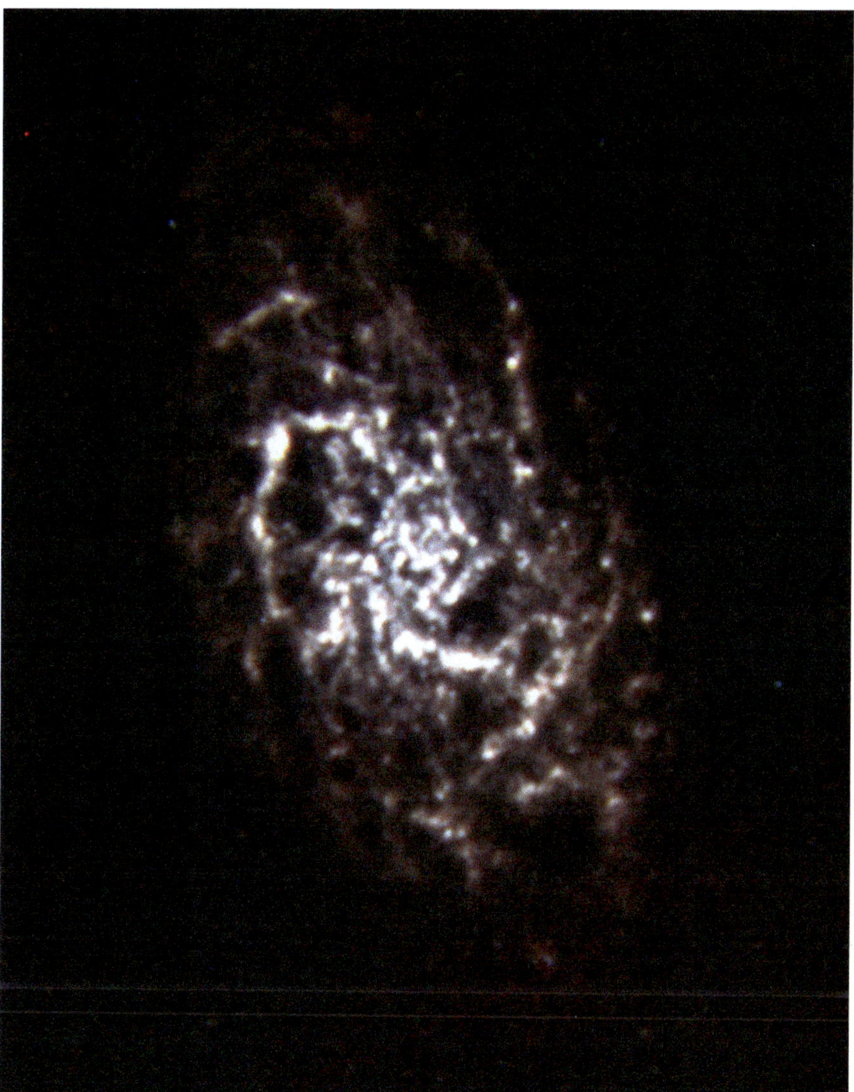

Fig. C.8 Far-infrared view of **M33** as obtained with the *HerschelSpace Observatory* at wavelengths of 250–500 μm. The emitting dust and associated molecular gas clouds trace out spiral arms, within which the giant HII regions are located (as seen in Fig. C.5). (Credits: ESA, *HerschelSpace Observatory*, GL Pilbratt et al. [2010], A&A, 518, L1 and MJ Griffin et al. 2010, A&A, 518, L3)

Fig. C.9 **M33** in the light of ionized hydrogen at 0.656-μm wavelength (in red), PAH molecules at 8- μm wavelength (in blue), and star-warmed dust at 24- μm wavelength (in green). (Credits: *Spitzer Space Telescope,* NASA, JPL-Caltech; from A. Weinbeck and W. H. Waller at https://galacticinquirer.net/2025/03/multi-spectral-imagery-of-the-multi-phase-ism-in-m33/)

C-3: Star-forming Spiral Galaxies beyond the Local Group

Within 100 million light-years of the Local Group, there are an estimated 160 other groups that together contain roughly 2500 giant galaxies. About half of these galaxies have spiral classifications, and about half again of these have nearly face-on orientations. This latter attribute enables the galaxy disks and their sundry contents to be optimally viewed. Recently, the *James Webb Space Telescope* was tasked to image a robust sample of these giant face-on spiral galaxies. The so-called Physics at High Angular resolution in Nearby GalaxieS (PHANGS) Collaboration has yielded astonishing images at near- and mid-infrared wavelengths (see Fig. C.10). These images have been complemented by corresponding images at radio wavelengths (with *ALMA*) and at optical wavelengths (with *HST*). The combined datasets have revealed galactic ecosystems in action across a broad range of barred and unbarred spiral galaxy morphologies.

In many of these galaxies, the spiral arms are accompanied by interarm regions of emitting dust and molecular gas clouds. Some of these emitting regions could represent future sites of star formation. Others might represent

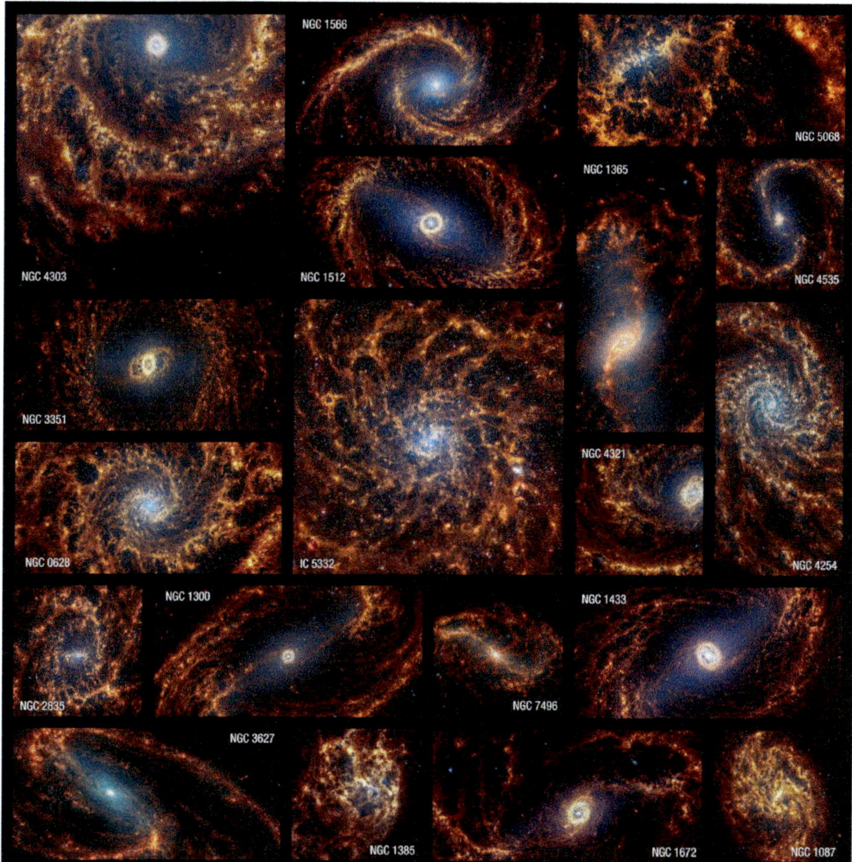

Fig. C.10 Gallery of giant spiral galaxies in the nearby Universe, as imaged in the near- and mid-infrared by *JWST*. The emitting dust and molecular gas clouds are seen to trace the spiral arms as well as other regions between the arms. In several of these galaxies cellular structure is evident (Credits: NASA/ESA *JWST* PHANGS Team, from Thomas G. Williams et al. [2024], *Astrophysical Journal Supplement Series*, Vol. 273, p. 13, https://iopscience.iop.org/article/10.3847/1538-4365/ad4be5)

vestiges of prior star-forming activity, after strong stellar winds and supernova blasts have had their way. I am struck especially by the cellular structure evident in some of the disks in these galaxies. The current poster child for this observed behavior is the Phantom Galaxy (aka M74, NGC 628), as shown in Fig. C.11. What has produced the sundry voids in the emitting matter? It could be the starburst-driven evacuation process previously mentioned. Or perhaps the shearing rotation within M74 and the other spiral galaxies imaged by *JWST* is shaping the observed ornate systems of gas and dust. For now, we can only speculate on the possibilities.

Fig. C.11 Multi-observatory views of the **Phantom Galaxy** (M74, NGC 628). *Left*—Visible light view from the *Hubble Space Telescope* showing giant HII regions and obscuring clouds of dust along the spiral arms. *Middle*—Blend of visible and infrared views from the *HST* and *JWST*. *Right*—Color-coded image of infrared emission, as imaged by the *JWST*. The bluer colors correspond to shorter wavelength infrared emission from PAH molecules, while the redder colors denote longer wavelength emission from star-warmed dust (Credits: ESA/Webb, NASA & CSA, J. Lee and the PHANGS-JWST Team. Acknowledgement: J. Schmidt)

Appendix D: Exemplary Starburst Galaxies

As the LMC, Triangulum Galaxy, and Phantom Galaxy attest, many irregular and spiral galaxies host starbursts as part of their overall makeup. Some galaxies, however, are completely overtaken by the starburst activity within them. They are typically called *starburst galaxies*. Most of these galaxies are relatively small, with irregular or amorphous classifications. In this appendix, we will consider the irregular starburst NGC 1569 and revisit the amorphous starburst M82 (NGC 3034). We will then ponder NGC 6240, a good example of starburst activity that has followed a merger of two gas-rich spiral galaxies. At high redshift, the starburst galaxies come in even more powerful versions, but are barely resolvable with our current telescopes. Their multi-wavelength properties, however, indicate that violent transformations were underway billions of years ago, when these galaxies were very young.

D-1: NGC 1569

The dwarf irregular galaxy, NGC 1569, is regarded as the closest starburst galaxy to us. Recent observations with the *Hubble Space Telescope*, have increased its estimated distance from 7.2 million light-years to 11.0 million light-years, making it part of the IC 342 galaxy group while (barely) retaining its status as the nearest starburst galaxy.

Perhaps because of interactions with other galaxies (or itinerant gas clouds) in this group, NGC 1569 is currently ablaze with starburst activity both within and beyond its stellar body (see Fig. D.1). Spanning only 16,000 ly, it nonetheless is forming stars at a rate close to that of the entire Milky Way Galaxy. These stars can be seen in the high-resolution image obtained with the *Hubble Space Telescope* (see Fig. D.2). Many of them are concentrated into two dense clusters that resemble NGC 3603 in the Milky Way Galaxy and R136 in the Large Magellanic Cloud. To have these dense and powerful star clusters operating within such a small galaxy evokes what it might have been like in the early Universe, when the first galaxies were emerging from the cosmic chaos.

The extended H-alpha emission suggests that NGC 1569 has been actively starbursting over a period of at least 30 million years. What we see today are

Fig. D.1 Ground-based visible view of the starburst galaxy, **NGC 1569**. Here, north is about 25-deg counterclockwise from the vertical. This image highlights the extensive outflows of H-alpha emitting gas. These outflows are fluorescing in response to the extreme UV light from the hot stars within the body of the galaxy. They extend more than 5,000 ly away from their stellar energizers. They were likely jettisoned by salvos of supernova activity over the past 30 million years (Credits: adapted from Adam Block/Josep Drudis/Mount Lemmon SkyCenter/University of Arizona, with permission)

Fig. D.2 High-resolution image of **NGC 1569**, as obtained with the *Hubble Space Telescope* and presented with a "natural" palette. The super star clusters A and B are located towards the center right of the galaxy. The EUV light and strong winds from the clusters have evacuated much of the gas near them. The dominant red emission is from ionized hydrogen at the working surfaces of the remnant clouds. (Credits: NASA, ESA, Hubble Heritage, STScI/AURA, A. Aloisi [ESA/STScI] et al.)

the detritus of prior supernova blasts, now energized by the current population of hot massive stars. Further support for this interpretation comes from observations at X-ray energies (see Fig. D.3). Here, we see a halo of million-degree gas co-extensive with the relatively cooler ionized hydrogen gas. The emitting coronal gas could be filling enormous cavities produced by the outflow, or alternatively, tracing the shock-heated interface between the evacuated interior and the outflowing gaseous medium. Velocities exceeding 100 km/s suggest that the outflow is escaping the galaxy, and therefore contributing to the intergalactic medium at the host galaxy's expense.

D-2: M82 (Reprise)

In Chap. 10, we first came to know M82 (NGC 3034) as a starburst galaxy, where most of the action is emanating from a circumnuclear region of dense molecular gas and consequent super star clusters. The resulting superwinds attest to the power generated when enough massive stars get formed in a concentrated region and subsequently expire in titanic supernova explosions. The bi-polar superwinds in M82 involve millions of Suns worth of matter

Fig. D.3 The nearby starburst galaxy **NGC 1569** in the light of optical H-alpha (coded red) and X-rays (coded green). The field of view is 4.5 arcmin by 4.5 arcmin, amounting to 14,300 ly by 14,300 ly at the galaxy's distance of 11 million light-years. North is up and east to the left. The X-ray emission comes from stellar point sources along with a diffuse halo that coincides with the diffuse H-alpha emission. The X-ray halo is thought to consist of million-degree gas that was shock heated by thousands of supernova blasts over the last 30 Myr. (Credits: NASA/UCSB/C. Martin et al. & NOAO/KPNO/C. Martin with reference to C.L. Martin, H.A. Kobulnicky, T.M. Heckman [2002] https://iopscience.iop.org/article/10.1086/341092/fulltext/)

shooting outward at thousands of km/s. The resulting mechanical power is equivalent to the radiant luminosity of 100 million Suns. This torrid outflow has carried with it a magnetic field of remarkable simplicity (see Fig. D.4).

Closer examination of these well-ordered fields has indicated that they are open rather than looping back to the galaxy. That means the superwinds are destined to escape the galaxy and so enrich the intergalactic medium with elements that were created as part of the supernova activity. The magnetic

Fig. D.4 Mapping of the magnetic fields associated with the bipolar superwinds in **M82.** The galaxy itself is shown in visible light as the grey "cigar" with obscuring dust lanes. The outflows are shown in the red color of optical H-alpha emission. The yellow traces infrared emission from warm dust as imaged by the *Spitzer Space Telescope.* The magnetic fields were mapped using the far-infrared imaging polarimeter aboard the *Stratospheric Observatory for Infrared Astronomy (SOFIA)* (Credits: NASA/ *SOFIA*/E. Lopez-Rodriguez/*Spitzer* /J. Moustakas et al.)

fields themselves would be free to contribute to the intergalactic medium's overall magnetism, thus setting the stage for magnetically-directed flows amongst the galaxies. The epoch of galaxy birth, some 12 billion years ago, likely involved similar dynamics, where ambient magnetic fields helped direct the accretion of matter onto new galactic systems.

As NGC 1569 and M82 attest, many starburst galaxies result from interactions with nearby objects. For NGC 1569, it could have been a run in with a large cloud of atomic gas that has fueled the starburst over the past 30 million years. For M82, the provocateur is much clearer, as a vast swath of atomic hydrogen envelops the giant spiral galaxy M81, linking it to both M82 and the peculiar elliptical galaxy NGC 3077 (see Fig. D.5).

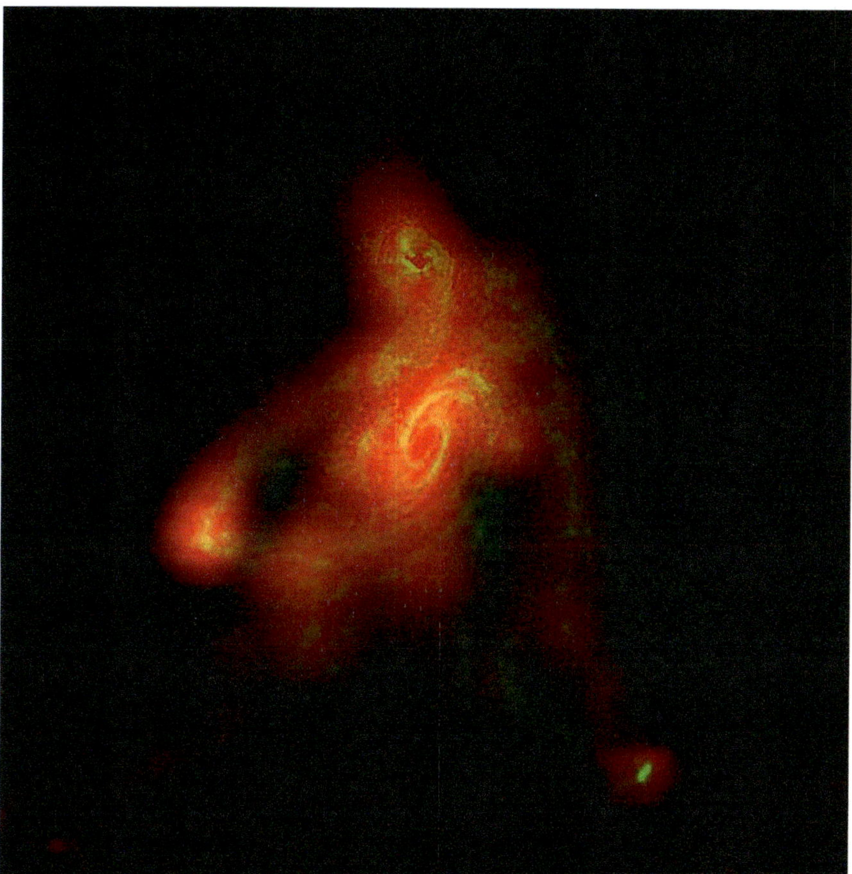

Fig. D.5 The **M81, M82, NGC 3077** system in the light of atomic hydrogen at a radio wavelength of 21 cm. The HI emission presented here is a combination of diffuse emission as mapped by the *Green Bank Telescope (GBT)* in red with more detailed emission from the *Very Large Array (VLA)* in green. The giant spiral M81 occupies the center, while M82 is located above it, and NGC 3077 is toward the left. The HI gas associated with M81 completely enfolds the other two galaxies (Credits: Chynoweth et al., NRAO/AUI/NSF, Digital Sky Survey)

D-3: NGC 6240

Other starburst galaxies result from direct mergers between two or more galaxies. NGC 6240 provides a relatively nearby exemplar of a starburst galaxy in the midst of the merging process (see Fig. D.6). Situated 400 million light-years in the constellation of Ophiuchus, the so-called Starfish galaxy is thought to consist of two spiral galaxies that have collided with one another. The gravitational mashup has led to a flamboyant distortion of the original spiral systems, with vast loops and tidal tails emanating away from ground zero.

Optical imaging has revealed 3 dense cores near the merger center (see inset in Fig. D6), while X-ray imaging has shown two of them emitting at these high energies (see Fig. D.7). The two X-ray sources likely indicate two supermassive black holes actively gorging on their surroundings. They also appear in the high-resolution mapping of molecular gas by *ALMA*. The third optical core is either an X-ray quiet black hole (suggesting a third galaxy has joined the morass) or a particularly active super star cluster.

Besides its evocative shape, NGC 6240 offers a relatively nearby view of an *Ultra Luminous Infrared Galaxy (ULIRG)*. These dusty systems radiate at luminosities equivalent to more than a trillion Suns—mostly in the infrared. For most of the observed ULIRGs, the relative contributions of luminosity from the starbursts and active galactic nuclei (AGN) remain unclear. That is because they are too dusty and distant to be well resolved. They are most prevalent at high redshifts, when the Universe was very young and much denser. Back then, the crowding of galaxies led to an enhanced frequency of mergers—with all the fireworks that ensued. We will explore that epoch next.

D-4: Starburst Galaxies in the Early Universe

When astronomers have used the *Hubble Space Telescope* and other powerful observatories to image as deeply as they can, they have found flurries of galaxies upon galaxies with redshifts and corresponding lookback times that take us back to the earliest epochs of galaxy evolution, a scant billion or so years after the hot big bang.. What they see are mere motes of light, what I like to call "little red turds at the edge of infinity" (see Fig. D.8).

One particular team of astronomers has spent the last 20 years assaying the amounts of gas and concurrent star-forming activity in these primordial systems. The Cosmic Evolution Survey (COSMOS) was designed to characterize the formation and evolution of galaxies along with their cosmic structuring. The survey spans a 2 square-degree field of view that has been imaged at multiple wavelengths spanning the electromagnetic spectrum. The team hit paydirt with observations at sub-millimeter and far-infrared wavelengths, as these observations respectively trace the amounts of star-forming gas and the resulting rates of massive star formation. The *ALMA* array of radio telescopes, observing at submillimeter wavelengths, provided well-resolved measures of the molecular gas masses in more than 700 galaxies contained within the COSMOS field of view. The galactic emission arises

Fig. D.6 The merging and starbursting system **NGC 6240**, as imaged by the *Hubble Space Telescope*. This peculiar system spans 300,000 ly and includes several tidal tails that contain recently-formed hot blue stars. The inset shows a zoomed view of the merger center, as obtained with the ESO's *Very Large Telescope (VLT)*. The northern source (N) traces a supermassive black hole, as revealed by its X-ray emission. The southern component (S) consists of another X-ray emitting black hole and a secondary source that is X-ray quiet. The green color indicates the distribution of gas ionized by radiation surrounding the black holes; the red lines show the contours of the starlight from the galaxy, and the length of the white bar corresponds to 1000 light years (Credits: P. Weilbacher, Leibniz Institute for Astrophysics Potsdam/ NASA/ESA/Hubble Heritage/STScI/AURA/Hubble Collaboration/A. Evans, University of Virginia, Charlottesville/NRAO/Stony Brook University, W. Kollatschny et al. 2020, https://doi.org/10.1051/0004-6361/201936540)

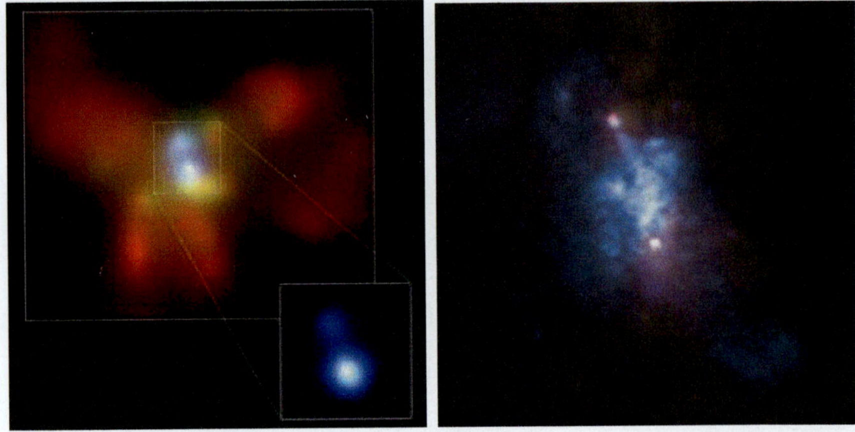

Fig. D.7 The starbursting core of **NGC 6240** at X-ray and radio wavelengths. *Left*—The X-ray image from the *Chandra Space Observatory* includes two bright sources that are thought to be supermassive black holes. These sources have a projected separation of about 4,500 light-years. The diffuse X-ray emission could arise from supernova activity associated with the nuclear starburst. *Right*—The higher-resolution radio image from *ALMA* distinctly shows the two SMBHs along with lots of CO-emitting molecular gas in between them. There could be more than 10 billion solar masses of molecular hydrogen contained within this concentrated region. (Credits: *Left*—NASA/CXC/MPE/S. Komossa et al., *Right*—ALMA/ESO/NAOJ/NRAO/E. Treister, https://iopscience.iop.org/article/10.3847/1538-4357/ab6b28)

from glowing dust in the gas clouds that was emitted at far-infrared wavelengths but has since been redshifted into the sub-millimeter. The scientists found that the long-wavelength "tail" of the submm-wave emission scales proportionately with the amount of dust and—by extrapolation—with the mass of gas in these galaxies (see Fig. D.9).

To gauge the rates of massive star formation, the COSMOS team used far-infrared observations from the European *Herschel Space Observatory* (see Fig. D.10). Here, the emission is dominated by glowing dust that has been warmed to higher temperatures by the newborn massive stars that are entrained within and among the clouds. Originating in the mid-infrared, the emission has since been redshifted into the far-infrared. The intensity of this emission directly scales with the number of short-lived massive stars, and by inference, the rate of massive star formation. We are now bearing witness to the most distant and powerful ULIRGs.

By tracking both the gas masses and starbirth rates in galaxies out to high redshift, the COSMOS team found that the masses and birth rates greatly exceeded those found in galaxies of the local (current-epoch) Universe by factors of 10-100. Moreover, the starbirth rates per unit mass of gas—ie. the efficiencies—were significantly elevated. It appears that galaxies in the early

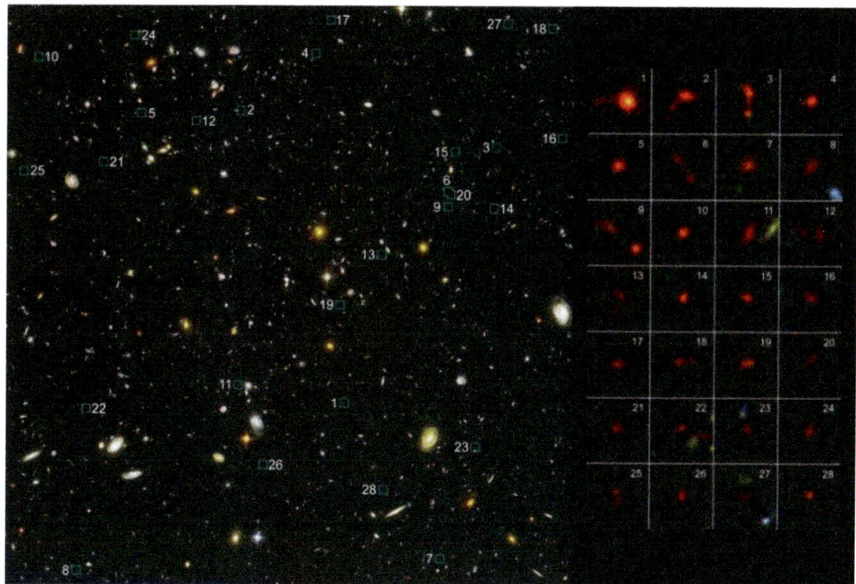

Fig. D.8 Between 2003 and 2014, the *Hubble Space Telescope* observed the same tiny region across 7 different wavelength bands. The resulting Hubble Ultra-Deep Field is dominated by galaxies near and far. Here, galaxies at highest redshift are identified on the left with green squares and shown at high magnification on the right (Credits: NASA, ESA, R. Bouwens and G. Illingworth [UC, Santa Cruz])

Fig. D.9 The *Atacama Large Millimeter/submillimeter wave Array (ALMA)* situated in the high Chilean desert, has observed more than 700 star-forming galaxies—assaying their molecular gas masses out to high redshifts and corresponding lookback times (Credit: European Southern Observatory [ESO])

Universe were truly ablaze with rampant starburst activity, just as they were assembling from the morass of sub-galactic clumps some 12 billion years ago. Throughout this book, I have referred to "galactic ecosystems" as the hosts of transformative processes involving the birth of stars and planets from clouds of gas. Here, the intensity of starburst activity in these primordial systems tempts me to use the term "galactic wackosystems" instead!

In his award lecture at the summer 2022 meeting of the American Astronomical Society, team leader, Nick Scoville attributed these tirades of massive star formation to collisions between molecular clouds during all that merging and melding. To find out how far back in time this transformational phase of galaxy evolution can be perceived, he and the COSMOS team anticipate that their targeted observing program with the *James Webb Space Telescope (JWST)* will provide yet new insights.

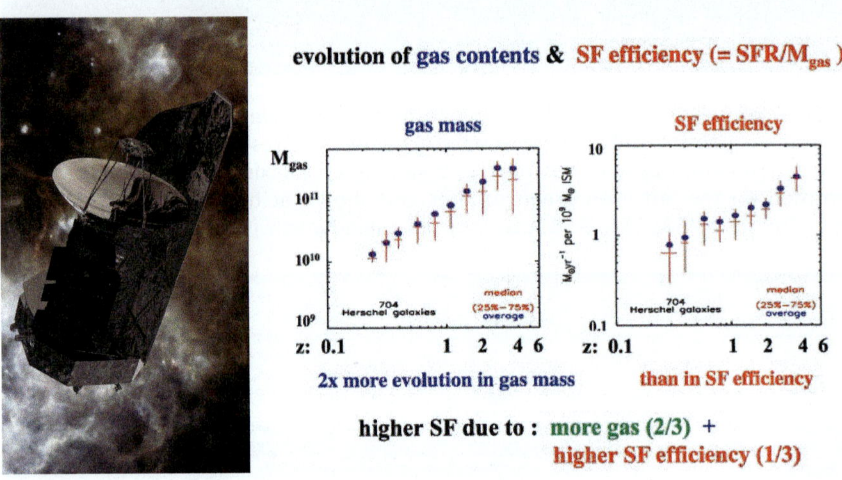

Fig. D.10 *Left*—Artist's depiction of the *HerschelSpace Observatory* and the star-forming nebulae that it senses in the far-infrared. *Right*—The gas mass plot shows that galaxies at high redshift (z) are laden with up to a trillion solar masses of star-forming molecular gas. The SF efficiency plot shows greatly enhanced rates of star formation per unit mass of molecular gas at the higher redshifts. A redshift of z = 4 corresponds to a lookback time of about 12 billion years (Credits: European Space Agency, N. Scoville and the COSMOS team)

Appendix E: Physical Approximations

In this golden age of revelatory observations and detailed numerical simulations, astronomers still rely on mathematical models of physical phenomena to make sense of what is going on. These models can help us ascertain the salient mechanisms behind the formation of stars and planetary systems – and of the consequential feedback processes impacting the natal clouds. Here, I have selected a few of these processes to mathematically model and so provide approximations of what to expect.

E-1: Gravitational Instabilities:

In his original derivation of the conditions for gravitational instability, Sir James Jeans (1877–1946) considered an infinite slab of gravitating matter that is perturbed by a sound wave. The resulting solution yielded a minimum wavelength, above which the matter would be unstable to gravitational collapse. This minimum wavelength is commonly known as the Jeans length.

We can achieve the same ends by simply considering the gravitational and kinetic energies that are involved. For gravitational collapse to occur, the gravitational binding energy of a particle must exceed its kinetic energy, ie.

$$-\text{G.E.} > \text{K.E.}$$

where the minus sign accounts for the gravitational energy being a negative quantity. For simplicity, we can consider the gravitational energy of a particle in a spherical cloud of mass (M) and radius (R), and assume that the kinetic energy is dominated by the thermal motions. In this case, the above inequality can be written as,

$$\frac{GMm}{R} > \left(\frac{3}{2}\right)\text{kT},$$

where m is the mass of the relevant particle. Rewriting this relation in terms of the density ρ (assumed constant throughout) yields,

$$\text{G}\rho\left(\frac{4\pi}{3}\right)\text{R}^2\text{m} > \left(\frac{3}{2}\right)\text{kT},$$

whose solution for the gravitationally unstable radius is

$$R > R_J = \left(\frac{9}{8\pi}\right)^{1/2}\left(\frac{kT}{G\varrho m}\right)^{1/2},$$

where R_J is the Jeans radius, the minimum radius above which the cloud is unstable to gravitational collapse. Working through the various constants (and using the mass of a hydrogen molecule for the relevant particle) yields

$$R_J(\text{cm}) = 1.5 \times 10^7\left(\frac{T}{\varrho}\right)^{1/2}$$

or

$$R_J(\text{ly}) = 1.6 \times 10^{-11}\left(\frac{T}{\varrho}\right)^{1/2}$$

where T is in Kelvins and ρ is in grams/cm^3. In terms of the more commonly noted particle number density (n), where ($\rho = n\ m_{H2}$), the Jeans radius becomes

$$R_J(\text{ly}) = 8.6\left(\frac{T}{n}\right)^{1/2}.$$

The corresponding Jeans mass is

$$M_J = \left(\frac{4\pi}{3}\right)R_J^3\rho,$$

which in terms of temperature and density becomes

$$M_J = \left(\frac{4\pi}{3}\right)\left(\frac{9}{8\pi}\right)^{3/2}\left(\frac{kT}{Gm[H_2]}\right)^{3/2}\rho^{-1/2}.$$

Doing the arithmetic yields

$$M_J(\text{grams}) = 1.4 \times 10^{22}T^{3/2}\rho^{-1/2}$$

or

$$\frac{M_J}{M(Sun)} = 6.9 \times 10^{-12}T^{3/2}\rho^{-1/2}.$$

In terms of the particle density (n), this reduces to

$$\frac{M_J}{M(Sun)} = 3.8\,\mathrm{T}^{3/2}\mathrm{n}^{-1/2}.$$

For a typical cloud filament, the number density is roughly 10^3 particles per cubic centimeter. That corresponds to an H_2 mass density of 3.34×10^{-21} g/cm^3. Combine that with a temperature of 10 Kelvins, and you get a Jeans radius of 0.9 ly and a Jeans mass of about 4 solar masses ... enough to make a few cloud cores or one core with multiple protostars incubating.

For a typical cloud core, the number density is roughly 10^5 particles per cubic centimeter. That corresponds to an H_2 mass density of 3.34×10^{-19} g/cm^3. Combine that with a temperature of 10 Kelvins, and you get a Jeans radius of 0.09 ly and a Jeans mass of 0.4 solar masses ... enough to make a low-mass star.

Plots of the Jeans radius and Jeans mass as a function of particle density and temperature can be found in Chap. 5.

E-2: Free-fall Timescales

Once gravity begins to dominate over all other resistive forces within a cloud core (or in the core of a star), it is worth asking how long it would take for the core to collapse in free-fall. One approximation is to equate the gravitational force (F_g) with the acceleration (a) of a particle according to Newton's 2nd law of motion, where

$$\mathrm{F_g} = \frac{GMm}{r^2} = \mathrm{ma},$$

with (G) being the constant of universal gravitation, (M) the mass of the collapsing core, and (m) the mass of the particle. This reduces to

$$\mathrm{a} = \frac{GM}{r^2}.$$

The radius of infall, in turn, is related to the acceleration and infall time (t_{FF}) by the kinematic expression

$$\mathrm{r} = \mathrm{R} = \frac{1}{2}\mathrm{a}t_{FF}^2,$$

$$\text{or R} = \frac{GMt_{FF}^2}{2R^2}$$

where (R) is the radius of the core. Solving for the infall time yields

$$t_{FF}^2 = \frac{2R^3}{GM}$$

which in terms of the average initial density, where M = (ϱ 4 π R^3)/3, becomes

$$t_{FF}^2 = \frac{6R^3}{G\varrho 4\pi R^3}.$$

This reduces to

$$t_{FF}^2 = \frac{3}{(2\pi G\varrho)}$$

so that

$$t_{FF} = \left(\frac{3}{2\pi G\varrho}\right)^{1/2}.$$

This result comes close to

$$t_{FF} = 0.7\left(\frac{1}{G\varrho}\right)^{1/2}.$$

Another derivation involving circular orbital motion as a proxy for free-fall obtains

$$t_{FF} = 0.8\left(\frac{1}{G\varrho}\right)^{1/2},$$

and a more thorough derivation that takes full account of the changing force and acceleration (L. Spitzer [1978], p. 267) produces

$$t_{FF} = 0.5\left(\frac{1}{G\varrho}\right)^{1/2}.$$

In terms of the particle number density (n), this solution becomes

$$t_{FF} = 0.5\left(\frac{1}{Gnm}\right)^{1/2},$$

which works out to

$$t_{FF} = 1.1 \times 10^{15}\left(\frac{1}{n}\right)^{1/2} \text{ s, or}$$

$$t_{FF} = 3.5 \times 10^{7}\left(\frac{1}{n}\right)^{1/2} \text{ years.}$$

A molecular cloud core with a high number density of n = 10^6 hydrogen molecules per cubic centimeter would collapse in free-fall over a timescale of 3.5×10^4 years. Lower densities of 10^5 and 10^3 molecules/cm^3 yield free-fall times of 1.1×10^5 and 1.1×10^6 years, respectively. The fact that many of these cloud cores can be found in molecular clouds says that something is keeping their population from collapsing in one huge star-forming burst. The clouds themselves are thought to have ages equivalent to the oldest stars that they have spawned, of order 10 million years. That means either new cores are continuously replacing the collapsed cores, or that several resistive agents are keeping the cores from rampantly collapsing. One such agent is the slow conversion of gravitational energy into thermal energy via the Kelvin-Helmholtz contraction process, as is presented next.

E-3: Gravitational Energies and Mediated Contraction Timescales

As a molecular cloud core continues to collapse to ever greater densities, it inevitably becomes opaque to its own thermal radiation. The source of opacity is the dust contained within the core. This trapped thermal energy resists the free-fall, producing a much slower contraction that tracks with the luminosity of the core. The luminosity (L), in turn, can be equated with the release of gravitational potential energy (E_G) according to the time derivative of the energy …

$$L = \frac{-dE_G}{dt},$$

The gravitational energy of the contracting core can be obtained by integrating over concentric shells, such that

$$E_G = -\int \left(\frac{GM[r]}{r}\right) dM,$$

where the limits of integration are from M = 0 to M = M (T.L. Swihart [1992], p. 120). Each concentric shell has a mass of

$$dM = 4\pi r^2 \rho dr,$$

and the mass of the core out to radius r is

$$M = \frac{4\pi r^3 \varrho}{3}$$

where the density (ρ) is an average for the core. Combining these expressions for M and dM into the integral yields a gravitational energy in terms of radius, such that

$$E_G = \left(\frac{16\pi^2 G\varrho^2}{3}\right) \int r^4 \, dr$$

where the limits of integration are from r = 0 to r = R. The integration produces

$$E_G = -\frac{16\pi^2 G\varrho^2 R^5}{15}$$

with the square of the average density being

$$\varrho^2 = \left(\frac{3M}{4\pi R^3}\right)^2 = \frac{9M^2}{16\pi^2 R^6}$$

which reduces the expression for the gravitational energy to

$$E_G = -\frac{9}{15}\left(\frac{GM^2}{R}\right)$$

or more simply

$$\mathrm{E_G} = -\frac{3}{5}\left(\frac{GM^2}{R}\right).$$

The Kelvin-Helmholtz timescale is then the time taken for the contracting core to release this gravitational energy at the observed luminosity which is

$$\mathrm{t_{KH}} = -\frac{E_G}{L}.$$

A handy example would be that of a contracting protostar with the luminosity of the Sun. The Kelvin-Helmoltz timescale for this protostar to contract would be

$$\mathrm{t_{KH}} = \frac{3}{5}\left(\frac{GM^2}{RL}\right)$$

which boils down to

$$\mathrm{t_{KH}} = 3/5\left(6.67 \times 10^{-8}\right)\left(2 \times 10^{33}\right)^2/\left(7 \times 10^{10}\right)\left(4 \times 10^{33}\right) = 5.7 \times 10^{14}\,\mathrm{s},$$

or 18.1 Myr for a Sun-like protostar. This timescale is much longer than the free-fall timescale, effectively preventing the prospect of runaway collapse. More comprehensive calculations pioneered in 1961 by Japanese astronomer Chushiro Hayashi lead to a longer timescale of 30 Myr for the mediated contraction, initial ignition of deuterium fusion, and culminating ignition of hydrogen fusion in a Sun-like young stellar object (YSO). The timescales for higher-mass protostars decrease roughly as the square of the mass. Evolutionary tracks of luminosity and temperature for a variety of YSO masses are presented in Fig. 5.3.

E-4: Sound Speeds and Evolutionary Timescales

A gaseous medium in thermal equilibrium will have a characteristic sound speed. Indeed, acoustic waves provide an important means for attaining and sustaining thermal equilibrium. For this reason, the sound speed sets a key limit on the timescale for a gaseous nebula to behave as one thermally coherent system. Conversely, the limiting timescales of other processes (e.g. the fusing lives of massive stars) along with the sound speed can set a limit on the size of such a thermally coherent system.

The sound speed itself can be readily derived by assuming an ideal gas and equating the mean kinetic energy of a particle to the corresponding thermal energy. For a gas, this becomes

$$\frac{1}{2}\, m_H v_S^2 = \frac{3}{2} kT,$$

where the pertinent mass here is m_H, the mass of the hydrogen atom, v_s is the average (root mean square) particle speed, k is Boltzmann's constant for relating temperature to thermal energy, and T is the temperature in Kelvins. The sound speed is taken as the average particle speed whose solution is

$$v_S = \left(\frac{3kT}{m_H}\right)^{1/2}.$$

Plugging in the values for k and m_H yields

$$v_S = 1.57 \times 10^4 T^{1/2} \text{cm/s, or}$$
$$v_S = 0.157\, T^{1/2} \text{km/s.}$$

In a molecular cloud at T = 10 K, v_S = 0.5 km/s (or 0.36 km/s if the mass of diatomic hydrogen is used), while in an HII region at T = 10^4 K, v_S = 15.7 km/s, and in a supernova remnant at T = 10^6 K, v_S = 157 km/s. In all these regions, the observed particle speeds significantly exceed the sonic speeds, thus indicating prevalent supersonic dynamics at play. But the sound speeds remain useful in terms of thermalization timescales. These depend on the size of the pertinent nebula, such that t = D/v_S. For a molecular cloud spanning 10 ly, the timescale becomes 6 × 10^6 yr. An HII region of similar size would equilibrate in 2 × 10^5 yr. And an equivalent supernova remnant would thermalize in only 2 × 10^4 yr.

In the absence of other more limiting timescales (e.g. gas consumption or expulsion timescales), the sonic timescale provides a handy measure of the thermalizing timescales, and more generally, of the evolutionary timescales t_{Evol} that characterize diverse galactic ecosystems. Further insights can be gained by letting the size of the GE vary. Then, the expression for the evolutionary timescale can be approximated by

$$t_{Evol} \sim 1\,\text{Myr}\,(D/15\,\text{ly})\left(10\,\text{km}\,\text{s}^{-1}/v_S\right).$$

According to this rule of thumb, the Orion Nebula, 30 Doradus, and the starburst galaxy NGC 1569 would then have evolutionary timescales of about 1 Myr, 20 Myr, and 300 Myr, respectively. The extended evolutionary times, in particular, open up the prospect of multiple star-forming events occurring within giant HII regions and larger starbursts. Indeed, the formation of star clusters and their disruptive consequences within large galactic ecosystems are far from coeval. A good example is 30 Doradus in the Large Magellanic Cloud, where regions of star formation with ages of < 1 Myr, 2–5 Myr, 3–10 Myr, and 10–20 Myr are evident across the 300 ly diameter starbursting field (E. K. Grebel and Y.-H. Chu [2000], AJ, 119, 787). Just as a large forest fire can host pockets of smaller fires at somewhat different times, so too large galactic ecosystems can involve several smaller non-coeval episodes of star-forming activity.

E-5: Modeling the Sizes of HII Regions—The Strömgren Sphere

In an HII region, the balance between the hydrogen photo-ionization rate and the recombination rate can be expressed as …

$$\mathrm{N_i = n_p n_e} \alpha_{\mathrm{eff}} \mathrm{V}$$

where N_i is the ionization rate produced by the hot massive star (in units of ionizations/s), n_p is the number density of protons (in units of #/cm^3), n_e is the number density of electrons (in the same units), α_{eff} is the effective recombination rate coefficient, and V is the volume of the photo-ionized region (in cm^3). Only electron recombinations to quantum levels of 2 or greater in the H-atom count, because a recombination to the ground state would emit an ionizing photon, thus nulling the recombination process. The reason why both the proton density and electron density are invoked is because "it takes two to tango." The effective H-atom recombination rate coefficient at a temperature of ~ 10^4 K has been estimated to be $\alpha_B = 2.6 \times 10^{-13}$ recombinations cm^3 / s. By assuming a spherical volume, $V = 4/3\ \pi\ R^3$, one can solve for the radius R in terms of everything else, whereby

$$\mathrm{R} = \left\{ \frac{3N_i}{(4\pi n_p n_e \alpha_{eff})} \right\}^{1/3}.$$

For a typical nebular density of 100 H-atoms/cm^3 and an H-ionizing rate of 10^{49} photons/s (typical of an O6-type star, one gets an HII region with a radius of about 10 ly, or a diameter of about 20 ly. A cluster containing multiple O-type stars will increase the radius by the number of such stars to the 1/3 power, thus yielding radii of 20, 50, and 100 ly for 10, 100, and 1000 stars, respectively. Any dust in the nebula will compete with the gas for the ionizing photons, thus reducing the size of the HII region. Current estimates of the dust co-mingling with the ionized gas suggest reductions in size by no more than a factor of two.

The above formulation for the size of a hydrogen cloud photo-ionized by an internal source was first proposed by Bengt Strömgren in 1937. Today, the so-called Strömgren Sphere continues to guide astrophysicists in understanding the nebular response to photo-ionizing sources (L. Spitzer 1978, AGGM Tielens 2005, N.S. Schulz 2005, B.T. Draine 2010).

E-6: Powering by Stellar Winds

Consider a stellar outflow with spherical symmetry and, at radius r, an imaginary shell in the outflow of differential thickness dr. The differential mass in the shell will equal the matter density, ρ(r), in the shell times the volume of the shell, such that …

$$dM = \rho(r) A\, dr,$$

where the area of the shell, $A = 4\,\pi\,r^2$, so that …

$$dM = \rho(r) 4\pi r^2 dr$$

Dividing both sides by the differential time, dt, then yields the mass-loss rate

$$\frac{dM}{dt} = \rho(r) 4\pi r^2 \frac{dr}{dt}$$

or more simply

$$\frac{dM}{dt} = \rho(r) 4\pi r^2 v,$$

where v is the velocity of the stellar outflow. These velocities range from several tens of km/s for red giant stars to several thousands of km/s for O-type and WR supergiant stars (T.A. Losinskaya 1992). These outflows convey mechanical power whose "luminosity" can be formulated from the common expression for kinetic energy, namely

$$K.E. = \frac{1}{2}Mv^2.$$

The corresponding mechanical luminosity of the stellar wind, L_W, is then the time derivative of the kinetic energy. By letting the velocity be constant, this ends up being

$$L_W = \frac{1}{2}\left(\frac{dM}{dt}\right)v^2.$$

These wind luminosities range from 1/10 the Sun's radiative luminosity for red giant stars up to 10^5 solar luminosities for O-type and WR supergiant stars (see Table 9.1).

E-7: Upper Limits of Stellar Mass and Luminosity—the Eddington Limits

Photons of light have energies and corresponding momenta. Upon interacting with matter, say in a star, these photons transfer their momenta over time and so exert a force. It is this outward force that can counteract the inward force of gravity and so drive an outflow. If strong enough, the so-called radiation pressure force can prevent any further accretion of matter, thus limiting the star's ultimate mass and luminosity. These limits are named after Sir Arthur Eddington who first formulated them in the 1920s. A fairly comprehensive treatment is provided by Hale Bradt (2008), but other decent derivations are available from Wikipedia and other sources on the internet.

Let's begin by considering a photon of energy, $E = h f$, where h is Planck's constant for quantum energetics, and f is the associated frequency. This can be re-written as $E = h c/\lambda$, where λ is the associated wavelength. According to de Broglie's relation (and associated work by Einstein), the momentum (p) of a particle can be described by $p = h/\lambda$. That means the photon's momentum and energy are related by $p = E/c$.

Upon interaction with matter, the photon's momentum will change over time and so produce a force on the matter, whereby

$$F_{ph} = \frac{\Delta p}{\Delta t} = \frac{E_{ph}}{(c\Delta t)}$$

Now, let's consider the flood of photons from the entire star. At the surface of the star, this flood produces a luminous flux (I), such that

$$I = \frac{L_R}{4\pi R^2}.$$

Because the radiative luminosity, $L_R = E_R / \Delta t$, the radiating energy over an arbitrary time interval (Δt) is $E_R = L_R \Delta t$. The total radiating energy per unit area then becomes

$$\frac{E_R}{A} = \frac{L_R \Delta t}{(4\pi R^2)}.$$

The net force per unit area (the radiation pressure, P_R) is then

$$P_R = \frac{E_R}{\{(c\Delta t)(4\pi R^2)\}} = \frac{L_R}{(c4\pi R^2)}.$$

This pressure is mediated by the interacting particle's effective cross-sectional area, σ, so that the total force on the particle is $F_R = \sigma P_R$, or

$$F_R = \frac{\sigma L_R}{(c4\pi R^2)}.$$

When this outward force equals the inward gravitational force, we get our limiting luminosity for stability, such that

$$\frac{\sigma L_R}{(c4\pi R^2)} = \frac{GMm_H}{R^2},$$

where we assume here that the gravitating particle is a hydrogen atom (or ion). The solution for the limiting luminosity is then

$$L_R = \frac{4\pi cGMm_H}{\sigma},$$

where we also assume the interaction cross-sectional area is that involving the hydrogen's electron, namely the Thompson cross-section, $\sigma_T = 6.65 \times 10^{-29}\ m^2 = 6.65 \times 10^{-25}\ cm^2$ In the c.g.s. system of units, this reduces to

$$L_R = 6.3 \times 10^4\ (M/grams)\ erg/s, \text{ or}$$

$$L_R = 1.26 \times 10^{38}\left(\frac{M}{M_{Sun}}\right) \text{ erg/s, so that}$$

$$\frac{L_R}{L_{Sun}} = 3.3 \times 10^4\left(\frac{M}{M_{Sun}}\right).$$

Now, the mass-luminosity relation for main-sequence stars more massive than 10 M_{Sun} goes approximately as

$$\frac{L_R}{L_{Sun}} = \left(\frac{M}{M_{Sun}}\right)^{3.0}, \text{ so that}$$

$$\left(\frac{M}{M_{Sun}}\right)^{3.0} = 3.3 \times 10^4\left(\frac{M}{M_{Sun}}\right), \text{ or}$$

$$\left(\frac{M}{M_{Sun}}\right)^{2.0} = 3.3 \times 10^4, \text{ which has the solution}$$

$\left(\frac{M}{M_{Sun}}\right) = 180$, for the Eddington Limit on stellar mass, and

$\frac{L_R}{L_{Sun}} = \left(\frac{M}{M_{Sun}}\right)^{3.0} = 5.8 \times 10^6$, for the Eddington Limit on stellar luminosity.

Smaller limiting masses and luminosities can be obtained, if the star has heavier elements in its atmosphere, whose many bound electrons provide an enhanced opacity to the emergent radiation. That is why the stars of higher "metallicity" that typically populate our Galaxy are thought to have more truncated upper limits to their masses and luminosities.[2]

Appendix F: Properties of Main Sequence Stars

Beginning with the quantitative comparison of apparent stellar brightnesses by Hipparchus in 150 BCE, and continuing with the spectral classifications of stellar types by Annie Jump Cannon and her fellow female "computers" at Harvard College Observatory, astronomers have made great strides

[2] A simpler approach to deriving the radiation pressure for the Eddington Limits is to find the equivalent energy density. That means dividing the radiant energy by the pertinent volume. The radiant energy is $E_R = L_R\ \Delta t$,. and the pertinent volume is the light propagation length times the surface area, $V = (c\ \Delta t)\ (4\ \pi\ R^2)$. Therefore, the energy density (and radiation pressure) is $P_R = L_R\ \Delta t\ /\ \{(c\ \Delta t)(4\ \pi\ R^2)\} = L_R\ /\ (c\ 4\ \pi\ R^2)$. Done!

Table F.1 Observed properties of nearby main-sequence Stars

Spectral type	Color (B–V)	Spectral features	Surface temperatures
O3 to O9	Violet-Blue (−0.33) to (−0.30)	Ionized helium	60,000 to 30,000 K
B0 to B9	Blue (−0.30) to (−0.06)	Neutral helium Hydrogen	26,500 to 10,000 K
A0 to A9	Blue-White (+0.00) to (+0.19)	Hydrogen (strongest)	9,900 to 7,800 K
F0 to F9	Yellow-White (+0.31) to (0.54)	Hydrogen (weak) Ionized Calcium	7,000 to 6,000 K
G0 to G9	Yellow (+0.59) to (+0.74)	Ionized Calcium Ionized iron	5,900 to 5,400 K
K0 to K9	Orange (+0.82) to (+1.35)	Ionized calcium (strongest) Sodium (strongest)	5,200 to 4,000 K
M0 to M9	Red (+1.41) to (+2.00)	Titanium oxide Calcium hydroxide	3,700 to 2,700 K

in assaying and understanding the salient properties of stars (Waller, 2013 and 2022). Herein, the spectroscopic properties of stars in their hydrogen-fusing Main Sequence phases are both plotted and tabulated (see Fig. F.1 and Table F.1).

The data in this table are based on Johnson, Harold L. (1966), *Annual Review of Astronomy and Astrophysics*, Vol. 4, pp. 193-206, the online encyclopedia http://en.wikipedia.org/wiki/Stellar_classification, and other sources. The spectral type runs from type-O for the hottest stars to type-M for the coolest stars. Within each spectral type, gradations of surface temperature are denoted by a number, with 0 being the hottest and 9 being the coolest of its type. The (B–V) color refers to the ratio of radiant fluxes from the stars as observed through blue (B) and yellow (V) filters. This ratio is expressed as a difference in magnitudes … the cooler the star's surface, the greater the magnitude difference and (B–V) color.

From the star's apparent brightness and distance, astronomers can determine its luminosity. And from the way the star dances with any stellar companion, they can surmise the star's mass. Enough mass-luminosity pairings have been assayed to show a clear relationship between the two properties, thus enabling astronomers to infer the stellar masses from just their luminosities. Computing the mass/luminosity ratio, in turn, is equivalent to dividing the available "fuel" by the "fire" and so deriving the star's total lifetime on the Main Sequence. This lifetime, when combined with the proportion of stars at birth, further yields the relative number of stars for each spectral type—as listed in Table F.2.

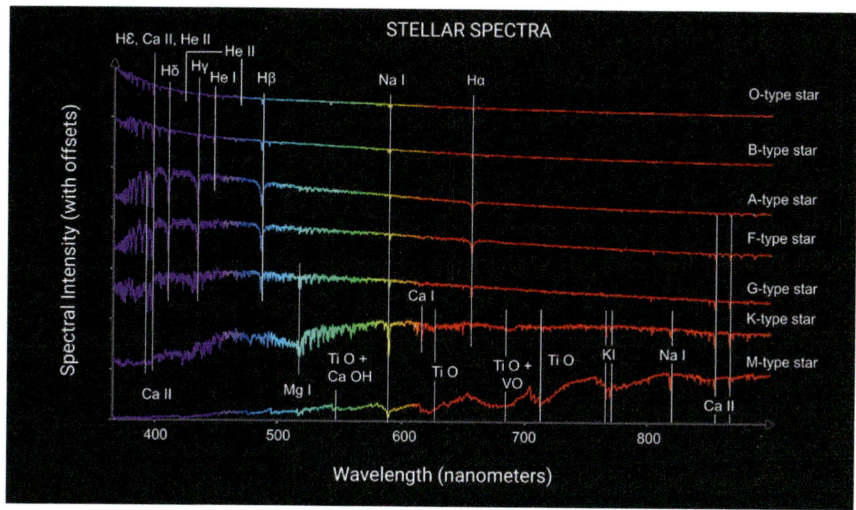

Fig. F.1 Sequence of representative stellar spectra for differing spectral types. The types trend from the hot violet-blue O-type stars at the top to the cool red M-type stars at the bottom. Different absorption-line features are evident within each spectral type, with the hydrogen (H) series of absorption lines being most prominent for the A-type stars and various molecular bands dominating the spectrum of the M-type stars. These absorption lines arise, as atoms and molecules in the stellar atmospheres consume light of particular wavelengths from the stellar interiors. The types of absorption lines depend critically on the surface temperatures of the corresponding stars. (Credits: Adapted by Leah Slingluff from the spectral sequence developed by the International Astronomical Union's Office of Astronomy for Education [IAU/OAE], Sloan Digital Sky Survey [SDSS], and Niall Deacon under Creative Commons CC BY-4.0 license)

Table F.2 Derived Physical Properties of Main-Sequence Stars

Spectral type	Mass	Luminosity	Lifetime	Relative number
O3 to O9	100 to 16 m_{Sun}	1,000,000 to 30,000 L_{Sun}	1.0 to 5.0 Myr	0.00003%
B0 to B9	16 to 2.1 m_{Sun}	30,000 to 25 L_{Sun}	5.0 to 840 Myr	0.13%
A0 to A9	2.1 to 1.4 m_{Sun}	25 to 5.0 L_{Sun}	0.84 to 2.8 Gyr	0.6%
F0 to F9	1.4 to 1.0 m_{Sun}	5.0 to 1.5 L_{Sun}	2.8 to 6.9 Gyr	3.0%
G0 to G9	1.0 to 0.8 m_{Sun}	1.5 to 0.6 L_{Sun}	6.9 to 13 Gyr	7.6%
K0 to K9	0.8 to 0.5 m_{Sun}	0.6 to 0.1 L_{Sun}	13 to 56 Gyr	12.1%
M0 to M9	0.5 to 0.1 m_{Sun}	< 0.1 L_{Sun}	> 56 Gyr	76.5%

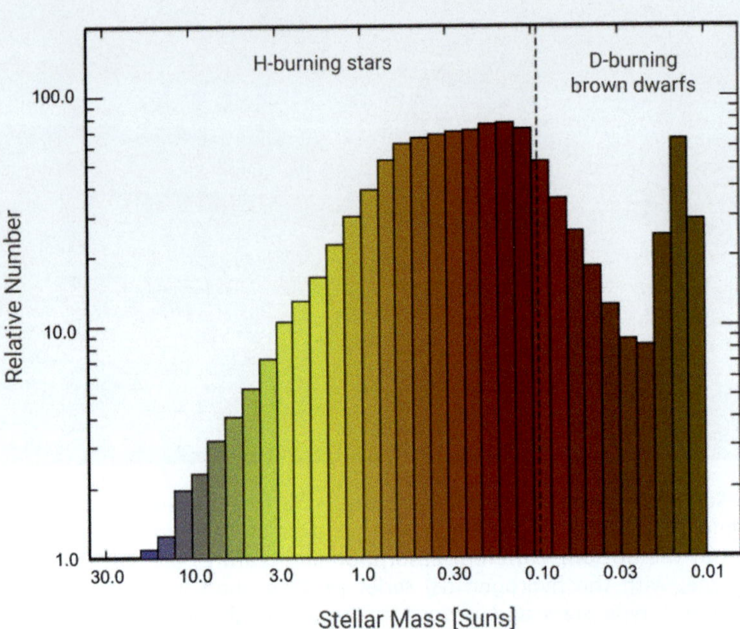

Fig. F.2 The distribution of stellar masses in a very young cluster is known as the Initial Mass Function, or IMF. This particular IMF was derived from the near-infrared luminosities of stars in the Trapezium Cluster—the closest substantial cluster to us that is young enough (~ 1 Myr) to still retain its hottest, most massive stars. The dotted line indicates the hydrogen fusing mass limit, below which deuterium-fusing brown dwarfs predominate. Above one solar mass, the stellar mass distribution can be described as a power law, where the relative number drastically decreases with increasing mass. This form of the IMF is consistent with the well-known Salpeter IMF. (Credits: Adapted from A. Muench, E. Lada, C. Lada, and J. Alves (2002), "The Luminosity and Mass Function of the Trapezium Cluster: From B Stars to the Deuterium-burning Limit," *Astrophysical Journal*, Vol. 573, pp. 366–393)

The masses, luminosities, and relative numbers in this table refer to values in http://en.wikipedia.org/wiki/Stellar_classification. Lifetimes are calculated from ($\tau = 10^{10}$ $(m/m_{Sun})/(L/L_{Sun})$ years). The relative numbers are derived from the distribution of masses at birth (the Initial Mass Function) combined with the relative lifetimes. A representative Initial Mass Function is shown in Figure F.2 below. This particular IMF refers to the Trapezium Cluster of stars that underlies and powers the Orion Nebula (M42). In the local field of our Galaxy, there are four B-type stars within 100 light-years of the Sun, while the nearest O-type star, Zeta Ophiuchi, is about 400 light-years away, consistent with their relative rarity.

Suggested Reading

General Interest

Bally J, Reipurth B (2006) *The Birth of Stars and Planets,* Cambridge University Press, Cambridge, UK

Barnard, EE (1927) *A Photographic Atlas of Selected Regions of the Milky Way*, eds. EB Frost and MR Calvert, Carnegie Institute of Washington, Washington, D.C., archival website at https://exhibit-archive.library.gatech.edu/barnard/

Cohen M (1988) *In Darkness Born: The Story of Star Formation*, Cambridge University Press, Cambridge, UK

Dick SJ (2019) *Classifying the Cosmos – How We Can Make Sense of the Celestial Landscape,* Springer Nature Switzerland, Cham, SUI

Fraknoi A, Morrison D, Wolff SC (2022 +) *Astronomy 2e,* OpenStax, Rice University, Houston, TX, https://openstax.org/details/books/astronomy-2e/

Kaler JB (1997) *Cosmic Clouds: Birth, Death, and Recycling in the Galaxy*, Scientific American Library, New York, NY

Powell R (2006) *Atlas of the Universe*, website at http://www.atlasoftheuniverse.com/

Waller WH (1985), "Mighty Winds from Newborn Stars," *Contact Magazine (U. Massachusetts),* vol. 9, p. 5

Waller WH (2013) *The Milky Way: An Insider's Guide*, Princeton University Press, Princeton, NJ

W. H. Waller, *Crucibles of Creation*, Astronomers' Universe,
https://doi.org/10.1007/978-3-032-17258-7

Waller WH (2021) "Sensing the Biochemical Character of Galactic Ecosystems," *The Galactic Inquirer,* http://galacticinquirer.net/2021/12/sensing-the-biochemical-character-of-galactic-ecosystems/

Waller WH (2022) *Astronomy: A Beginner's Guide,* OneWorld Publishers, London, UK

Wynn-Williams G (1993) *The Fullness of Space,* Cambridge University Press, Cambridge, UK

Technical

Anderson LD et al. (2014) "The WISE Catalog of Galactic HII Regions," *Astrophysical Journal Supplement,* 212: 1-18, interactive website at http://astro.phys.wvu.edu/wise/

André P et al. (2019) "Probing the Cold Magnetized Universe with SPICA-POL (B-BOP)" *Publications of the Astronomical Society of Australia (PASA), 36, e029,* https://doi.org/10.1017/pasa.2019.20

Andrews SM (2020) "Observations of Protoplanetary Disk Structures," *Annual Review of Astronomy and Astrophysics,* 58:483-528

Bradt H (2008) *Astrophysical Processes – The Physics of Astronomical Phenomena,* Cambridge University Press, Cambridge, UK

Donagal-Goldman, SD & KE Wright (2021) *The Astrobiology Primer v. 2.0,* Mary Ann Liebert, Inc. publishers, https://www.liebertpub.com/doi/10.1089/ast.2015.1460.

Draine, BT (2010) *Physics of the Interstellar and Intergalactic Medium,* Princeton University Press, Princeton, NJ

Evans, NJ II (1999) "Physical Conditions in Regions of Star Formation," *Annual Review of Astronomy and Astrophysics,* 37:311-362

Feitzinger JV, Stuewe JA (1984) "Catalogue of Dark Nebulae and Globules for Longitudes 240 to 360 Degrees," *Astronomy & Astrophysics Supplement Series,* 58:365-386

Fontani F (2024) "Observations of Phosphorus-bearing molecules in the interstellar medium," *Frontiers of Astronomy & Space Science,* 11: http://dx.doi.org/10.3389/fspas.2024.1451127

Grebel, EK, Chu Y-H (2000), "*Hubble Space Telescope* Photometry of Hodge 301: An "Old" Star Cluster in 30 Doradus," *Astronomical Journal*, 119: 787, https://iopscience.iop.org/article/10.1086/301218/meta

Heyer M, Dame TM (2015) "Molecular Clouds in the Milky Way," *Annual Review of Astronomy and Astrophysics,* 53: 583-629, https://doi.org/10.1146/annurev-astro-082214-122324

Jorgensen JK, Belloche A, Garrod RT (2020) "Astrochemistry during the Formation of Stars," *Annual Review of Astronomy and Astrophysics,* 58:727-778, https://doi.org/10.1146/annurev-astro-032620-021927

Kolb, VM, 2019, *Handbook of Astrobiology*, CRC Press, 2021.

Laques P and Vidal JL (1979) "Detection of a New Kind of Condensations in the Center of the Orion Nebula, by Means of S20 Photochathodes Associated with a Lallemond Electronic Camera," *Astronomy & Astrophysics,* 73: 97-106

Losinskaya TA (1992) *Supernovae and Stellar Wind in the Interstellar Medium,* American Institute of Physics, New York, NY

Lynds BT (1962) "Catalogue of Dark Nebulae," *Astrophysical Journal Supplement,* 7:1-52, https://articles.adsabs.harvard.edu/pdf/1962ApJS....7....1L

Öberg, KI (2016), "Photochemistry and astrochemistry: photochemical pathways to interstellar complex organic molecules," in *Chemical Reviews,* 2016, https://arxiv.org/pdf/1609.03112.pdf.

Öberg KI, Fachini S, Anderson D (2023), "Protoplanetary Disk Chemistry," *Annual Review of Astronomy and Astrophysics,* 61:287-328

Pineda J, Goldsmith PF, Chapman N, et al. (2010) "The Relation of Gas and Dust in the Taurus Molecular Cloud," *Astrophysical Journal,* 721: 686

Rathborne JH et al. (2009) "Dense Cores in the Pipe Nebula: An Improved Core Mass Function," *Astrophysical Journal,* 600: 742, https://iopscience.iop.org/article/10.1088/0004-637X/699/1/742

Reipurth B (ed) (2008a) *Handbook of Star Forming Regions, Vol. 1: The Northern Sky,* ASP Monograph Series, https://www.aspbooks.org/a/volumes/table_of_contents/?book_id=610

Reipurth B (ed) (2008b) *Handbook of Star Forming Regions, Vol. 2: The Southern Sky,* ASP Monograph Series, https://www.aspbooks.org/a/volumes/table_of_contents/?book_id=611

Sahu D, Liu S-Y, Liu T (2021) "Anatomy of Orion Molecular Clouds – The Astrochemistry Perspective/Approach," *Frontiers of Astronomy and Space Science,* https://doi.org/10.3389/fspas.2021.672893

Schulz NS (2005) *From Dust to Stars: Studies of the Formation and Early Evolution of Stars,* Praxis Publishing-Springer, Chichester, UK

Shaw AM (2022) *Astrochemistry: The Physical Chemistry of the Universe,* 2nd Edition, John Wiley & Sons, Hoboken, NJ

Shields GA (1990) "Extragalactic HII Regions," *Annual Review of Astronomy and Astrophysics,* 28:525-560

Spitzer L Jr. (1978) *Physical Processes in the Interstellar Medium,* Wiley-Interscience, New York, USA

Swihart TL (1992) *Quantitative Astronomy: Topics in Astrophysics,* Prentice-Hall, Inc., Englewood Cliffs, NJ, USA

Tielens, AGGM (2005) *The Physics and Chemistry of the Interstellar Medium,* Cambridge University Press, Cambridge, UK

van Dishoeck, EF et al. (2011) "Water in Star-forming Regions with the Herschel Space Observatory (WISH): Overview of key program and first results." *Publications of the Astronomical Society of the Pacific*, 123: 138

Veilleux S, Cecil G, Bland-Hawthorn J (2005) "Galactic Winds," *Annual Review of Astronomy and Astrophysics,* 43: 769-826

Waller, WH et al. (1996) *Fine-Scale Structure in the Far-Infrared Milky Way,* Final Report on NASA Contract #NAS5-32591 to StarStuff Incorporated, https://ntrs.nasa.gov/archive/nasa/casi.ntrs.nasa.gov/19960008844.pdf.

Waller, WH, Varosi F, Boulanger F, Digel, SW (1998) "Survey of Fine-Scale Structure in the Far-Infrared Milky Way," in *New Horizons from Multi- Wavelength Sky Surveys,* IAU Symp. 179, eds. B J. McLean, DA Golombek, J E Hayes, HE Payne (Dordrecht: Kluwer), p. 194, https://arxiv.org/pdf/astro-ph/9612233

Zhang, M (2023) "Distances to Nearby Molecular Clouds Traced by Young Stars," *Astrophysical Journal Supplement Series,* 265: 59–82

Index

W. H. Waller, *Crucibles of Creation*, Astronomers' Universe, https://doi.org/10.1007/978-3-032-17258-7